kitchen knife skills

kitchen knife skills

Techniques for carving, boning, slicing, chopping, dicing, mincing, filleting

Marianne Lumb

APPLE

A QUARTO BOOK

First published by Apple Press in the UK in 2009
7 Greenland Street
London NW1 0ND
www.apple-press.com

Reprinted 2010 (three times), 2012

Copyright © 2009 Quarto Publishing plc

ISBN: 978-1-84543-334-5

QUAR.NKS

Conceived, designed and produced by
Quarto Publishing plc
The Old Brewery
6 Blundell Street
London N7 9BH

QUAR: NKS

Project Editor: Emma Poulter
Art Editor: Emma Clayton
Designer: Karin Skånberg
Copy Editor: Jenny Doubt
Photographer: Andrew Atkinson
Proofreader: Claire Waite Brown
Indexer: Dorothy Frame
Art Director: Caroline Guest

Creative Director: Moira Clinch
Publisher: Paul Carslake

Colour separation by PICA Digital Pte Ltd, Singapore
Printed by Star Standard Industries (PTE) Ltd, Singapore

10 9 8 7 6 5

Contents

Foreword

I grew up watching my father
sharpening his knives. He was a
butcher and I remember standing
in the butcher's shop absolutely
transfixed at the sight of him
sharpening them very quickly.
Training as a chef in the heat of the
kitchen, my passion for knives and
their potential developed rapidly,
and in this book I share the skills I
have learned.

So what makes good knife skills?
Practice, dedication and learning from
experts. For me, the most enjoyable
part of cooking is the actual cooking
action itself – the shaking of the
pan, the flambé, the alchemy. But
preparation is key, and at the heart
of preparation and beautiful results
are sharp knives and well-executed
knife skills. This book encourages the
development of knife skills, and will
prove a useful guide to mastering a
professional finish at home.

MARIANNE LUMB

About this book

Kitchen Knife Skills **is a comprehensive
guide to choosing the right knife and
correct method of preparation for a
number of different types of food.**

Introduction (pages 8–35)

Beginning with an introduction to the subject of knives –
safe working practices, the anatomy of a knife, types of knives
available and sharpening methods – this section also includes
information on other cutting implements, chopping boards and
step-by-step instructions on how to implement core cutting
methods referenced throughout the book.

Techniques (pages 40–171)

Organised into chapters on food group (Vegetables and herbs, Fruit, Meat and poultry, Fish and seafood and Bread, pastries and cake), the bulk of the content provides step-by-step photographs and instructions on how to prepare every food type. With a host of useful information, tips and cross-references, this section will help you master the skills necessary to achieve professional results.

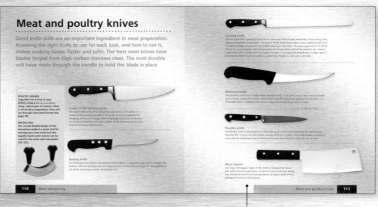

Meat and poultry knives

Good knife skills are an important ingredient in meat preparation. Knowing the right knife to use for each task, and how to use it, makes cooking easier, faster and safer. The best meat knives have blades forged from high-carbon stainless steel. The most durable will have rivets through the handle to hold the blade in place.

110   Meat and poultry

Meat and poultry knives   111

At the beginning of each chapter, the key knives useful for the preparation of that particular food group are identified.

To help you get the best results, the most suitable knife for each technique is identified, and alternatives suggested where appropriate.

You'll find a range of useful information, tips and suggestions on the knife skills being demonstrated, the knives themselves, or the foods being prepared.

Clear and concise technical instructions show you how to implement each of the knife skills.

Onions

Onions serve to flavour a variety of foods, and as one of the most widely used vegetables in cooking, learning how to cut an onion is an essential knife skill.

To get the best quality out of your onions, they should be cut just before you are going to use them. Try to cut the onions in a uniform manner and avoid 'cross chopping' the onion (see page 34). This produces inconsistent results and increases the gaseous aroma.

ONION KNIFE?
Slicing and chopping

Use a medium-sized cook's knife for both of these tasks. For larger varieties, such as Spanish onions, you may want to use a large cook's knife. For shallots, use a small cook's knife or kitchen knife.

The gaseous substance allyl sulphide is released when an onion is chopped, and it is this chemical that makes the eyes water. Many cooks have their own tested method of avoiding these tears, including wearing goggles. Try freezing the onion for 10 minutes before preparation. When the onion is cold, little aroma is given off, thus reducing its effect.

Chopping shallots

Similar in shape, the shallot differs slightly from the onion because of the interior bulbs that make up the vegetable.

Peel, chop and handle shallots in the same way as onions, keeping the root intact during preparation.

SLICING

Step 1
Grip the onion in one hand and steady it. Cut off the papery end of the onion to produce a flat base.

Step 2
Stand the onion up on its flat base. Cut in half, straight through the root. Aim to go directly through the centre of the onion, to produce two equal halves.

Step 3
Peel each of the onion halves, trimming the root ends but leaving the root intact. This will make it easier to maintain a regular-sized slice. Place both halves flat-side down onto your chopping board.

Step 4
Secure the onion using the claw method (see page 32). Steady the tip of your knife on the board and use a rolling chop (see page 32), to slice through the onion. Make sure you use the full length of your knife, in a forward cutting motion.

DICING

Step 1
Take a peeled, halved onion with the root still intact (as in step 3 above). Make cuts lengthways through the onion, using the tip of a cook's knife. Take care not to cut right through the root end of the onion, but ensure that the knife cuts to the bottom of the onion (i.e. through to the board).

Step 2
Continue to make cuts lengthways all the way across the width of the onion. As you begin to make your way across, change to the under-the-bridge method of cutting (see page 30). Bear in mind, the closer the cuts are together, the finer the end result. For larger dice, make the incisions further apart.

Step 3
Next, make slices as a horizontal cut into the onion as shown. Start from the bottom of the onion. The number of cuts you make depends on how small you want the dice to be. For larger dice you make fewer cuts, spaced further apart.

Step 4
Now cut vertically down using the rolling chop (remember to use the entire length of your knife). Be sure to make definite cuts throughout the onion cutting process. You can cross chop at the end if you wish too, but this increases the unpleasant gas aroma from the onion.

40   Vegetables and herbs

Onions   41

Safe working practices

Whether you are preparing food, or cleaning, sharpening or transporting your knives, these sharp implements must be treated with the utmost of care at all times.

Choosing the appropriate knife for your intended task and following the safe working practices shown here, you will be able to use your knives in the safest possible way.

Use a good knife

Whatever your method of preparation, always use a good quality knife. A good knife will have a sharp machine- or hand-ground blade, a comfortable handle and a deep-forged shoulder or finger guard. A poorly made knife will not take a sharp enough edge to cut efficiently, making the intended task much more difficult and more likely to become dangerous.

First aid

When using sharp tools regularly, you may at some point cut yourself, no matter how careful you are. Invest in a fully stocked first-aid kit to deal with minor injuries. If you cut yourself badly, dress the wound as best you can and go straight to Accident and Emergency for proper treatment.

Similarly, using the correct knife for the job will result in a more efficient performance.

Safe storage

Slotted knife blocks are the safest way to store knives, although these can be bulky and clutter a kitchen surface. Instead, try attaching a slotted knife rack to the wall. This will save on surface space and ensure knives are kept out of reach of children. Magnetic wall strips are also a great way of keeping knives safe and easily visible and accessible during cooking.

Never store knives in a drawer, unless you have a proper knife-storage system. The blades (and unsuspecting fingers) are easily damaged if knives are left to roam free in drawers. If you have to store in a drawer, be sure to keep the knives in the box they came in or wrapped in a sufficient protective covering.

Knife clips

Plastic clips are available that simply clip onto and cover the sharp side of the knife. These are great for even safer storage and travel.

Transport

In the event that you need to transport your knives, safety should be of primary concern. For professionals, knife rolls and cases are the best way to store knives during transport. Always maintain

Knife blocks come in a range of shapes and sizes, and also come with a slot to house a steel.

knife rolls well – over time holes may appear and knives may protrude or slip through.

If travelling abroad, knives have to go in the 'hold' in airplanes as they are potentially offensive weapons. Always seek advice from the appropriate authority before attempting to travel to ensure the correct protocol and regulations are being followed.

Taking care of your knife

Good knives will last a lifetime if you look after them. Methods of caring for your knives are covered in greater detail later in the book, although the condition of your knife can have an impact on safety, and so it is worth mentioning two key 'care' points here:

Always sharpen knives regularly
A blunt or dulled knife will not make a sharp or clean cut, and the desperate effort involved in trying to make such a knife perform can be dangerous, as well as frustrating and time consuming. A blunt knife is much more likely to slip and cause injury than a perfectly sharpened knife. (For more on sharpening knives, see pages 24–29).

Keep your knives clean A wet or greasy blade will not adhere to a magnetic wall strip and may also cause the knife to slip when cutting. Some knives may be dishwasher safe, but the heat of the dishwasher can affect the tempering, so should be avoided as a rule. Instead, hand wash knives separately with detergent and warm water, before rinsing well and drying with a towel. Soaking in a sink is not good for either the blade or the handle, and so should also be avoided.

For a handy and safe way of transporting your knives, use a knife roll.

As well as the health and safety concerns already referenced, here are some key points to help you use your knives safely and get the best possible results.

- Choose an appropriate knife for the task: using the correct knife for the job will mean you can make your cut as efficiently and effortlessly as possible.

- Always walk with the knife down by your side – don't wave the knife around like a sword.

- Always wear closed shoes. In the summer it may be tempting to wear open-toed shoes in the kitchen, but knives may fall onto the floor and cause injury.

- Keep pets and small children out of the way.

- Be sure to keep ingredients as cold as possible, particularly meat, poultry and fish. This is not only more hygienic but also ensures ingredients are as firm as possible, making them easier to cut.

- Put a cloth underneath your chopping board to steady it when working.

- Get organized and work in a tidy, methodical manner. Finish one task and place ingredients in the fridge before starting another job. Keep a dishcloth to hand at all times and clean your surfaces regularly.

- Keep your knives in top condition: sharpen either before or after every use, as automatically as you would wash them.

- Avoid wet or soiled hands. You need to maintain a steady grip of the knife's handle if you are to remain in complete control.

- Never leave knives in the sink, especially if it is full of hot soapy water. As well as having a detrimental effect on the blade and handle of a knife, it is all too easy to reach into a sink full of soapy water and inadvertently grab a knife by the blade.

- Be confident, but always alert – you are using a sharp implement!

Anatomy of a knife

Understanding the components and key features of a good knife will mean you are able to make the best of this tool. Depending on the variety of knife you are using, these components may differ between types and manufacturer, although the basic anatomical parts will be the same.

Whatever you choose, a good knife should feel well balanced and sturdy in your hand, have a comfortable handle and a deep-forged shoulder or finger guard. Today's ergonomic design means that modern knives can be comfortable for a chef to hold for up to 16 hours a day.

Invest in a decent set of knives and they will last you a lifetime. Always buy from a trusted shop or a well-known manufacturer with a good reputation.

Blade
This is the main body of the knife. Depending on the type of knife and its intended function, the blade will be either flexible or rigid. Fingertips and thumb, or the heel of your hand, can be placed on the top of the blade for extra control during cross-chopping (see page 34).

Point
At the very tip of the blade, the point of the knife is often used for scoring and piercing films, and is also used during the under-the-bridge cut (see page 30). The point is an extremely important feature of the boning knife, since it is used in a tip-first, dagger-like motion (see page 35).

Heel

The 'heel' refers to the base of the blade; the part primarily used for rolling chops (see page 32). This part of the knife is very sturdy and probably the easiest part of the knife to use: it's closest to the hand and so a good balance is easily achieved.

Bolster

Generally curved, this is the thick piece of metal between the handle and the blade that provides weight and balance.

Handle

Handles can be made of plastic, wood and sometimes bamboo or poly materials. Whatever the material, a handle should be comfortable to hold for a sustained period of time, and enable a good grip and control. Some knives may look glamorous but will cause calluses at the base of your fingers if held for long lengths of time. Generally speaking, you should hold the handle of the knife like you are shaking someone's hand (see *Core Cutting Methods*, pages 30–35 for exceptions to this rule).

Cutting edge/belly of the knife

Arguably the most important feature, this is the part of the knife that is sharpened. The cutting edge differs from knife to knife: depending on the type of knife, it can be smooth, scalloped, serrated or fluted, or have a hollow edge. Some Japanese knives may be beveled on one side only. The knife's edge should be sharpened regularly if it is to perform to the best of its ability. (For more on sharpening knives, see pages 24–29.)

Tang

The 'tang' refers to the upper end of the blade that fits into the handle. On all knives it should extend well into the handle. Tuck fingers behind the tang to protect them from the cutting action. On some knives, the tang ensures the knife will not lie flat on a surface.

Rivets

The rivets attach the tang (upper end) of the blade to the handle. These should be totally flat so they don't harbour any food or germs. Avoid knives in which the handle and blade are held together with glue instead of rivets – these are usually of a poor quality. The most modern type of knives are, however, forged from a single piece of metal and so do not have rivets. They are full-tang in design – that is, the blade extends the full length of the handle. Alternatively the blade may be moulded to the handle.

Types of knives

With climate, local produce, religion and lifestyle differing around the globe, world cuisine is extremely diverse and so are the tools with which to prepare it. Specific knives are designed for specific tasks, although given the advance in all areas of the international food industry, there is an overwhelming variety of knives on the market today.

Knives are generally made from mixes of various blends of stainless steel, ceramic and titanium, although there are new mixes and variations being created all the time.

Japanese knives are the smoothest and sharpest knives – the Japanese rarely use serrated knives (if ever) – and the best are handmade by craftsmen who train for at least 20 years. The Japanese are masters of all things ceramic and are the main producers of ceramic knives. Western knives are mainly made from metal blends in Germany, Switzerland and Sweden, primarily by machine.

Ceramic and titanium knives

PROS These knives are extremely sharp and tough. Even though they are strong they are lightweight and some chefs swear by them. These knives keep their edge for longer and so don't need sharpening as much as regular steel knives. Also, an apple cut with a ceramic knife won't brown as easily as that cut with a steel one.

CONS There is no 'give' in the material at all – ceramic knives are brittle and can chip. Sharpening these knives is inconvenient and expensive.

Steel knives

PROS The most traditional knife material, steel makes for durable and robust knives. Once mastered, it is easy to maintain sharpness.

CONS These knives are high maintenance – they lose their edge quickly and require regular sharpening.

Choosing the right knife for the job

As with any situation, a task is much easier to complete effectively if you are using the appropriate tool. Food preparation is no different, and so whether you are preparing fruit, vegetables, meat, poultry or fish, it is important to use the correct knife. A boning knife, for example, is specifically designed to separate meat from the bone easily – its shape is suited to the dagger-like motion with which it should be used. Using a cook's knife in this way will not only prove more difficult, but simply won't be as effective.

Fundamentally, a knife should go where you want it to go. When selecting your knife, this should be your primary consideration. There are three key questions to ask yourself when trying to identify the right knife for your task:

1. What is the knife for?
2. What size knife do you need, and are you going to use the full length of the blade?
3. What feels comfortable for you? Try holding the knife before you buy it.

Use the following directory to help you identify the perfect knife for your task. It is worth pointing out that the names of knives may vary between manufacturers.

Blade edges

As well as the various materials from which they can be made, knives are also available with different blade edges suited to certain ingredients and/or cutting movements.

Straight edge

This type of blade allows a smooth, clean and precise cut. Some knives have an edge on one side only, making for an even sharper blade. With such cases, left-handers will need to check they purchase a knife edged on the correct side. Ask your supplier.

Serrated edge

A serrated blade has a cutting edge comprising of a series of indentations along its length. The indentations mean it has less contact with the surface area being cut, and so the applied force at each point of contact is greater. Serrated blades are best for cutting through tough crusts and handling delicate interiors.

Scalloped edge

This edge is basically a mirror image of a serrated blade. The curves of a scalloped knife are inverted offering gentle but effective slicing. Perfect for cutting through delicate pastry, this knife also works well with slightly more rigid exteriors.

Tip

When purchasing a hollowed or fluted knife, examine the indentations – these should be significant if they are to be effective.

Hollowed or fluted edge

Indentations on the side of this type of blade create pockets of air that prevent extra thin, soft or sticky slices from sticking to the blade. Such blades are ideal when preparing starchy, fatty or oily ingredients.

General everyday knives

Your personal collection of knives should be tailored to suit your needs as a cook. General everyday knives are those that you reach for automatically for most food preparation, starting with chopping an onion or crushing garlic, for example. Above all, these knives should be comfortable for you. Instead of buying a large set, start by investing in a few very good ones (see *The Essential Eight*, page 17). Unless the blade is serrated, keep a steel at hand to top up the sharp edge, when necessary.

Cook's knife

An essential knife to have, this all-purpose knife works well with most foods and with most cutting methods (particularly the rolling chop, see page 32) and comes in small, medium and large. The sturdy spine of the knife (the top side of the blade that isn't sharp) can be used when breaking shellfish. As the main body of the knife is flat, it can also be used for flattening out a soft ingredient such as a veal escalope.

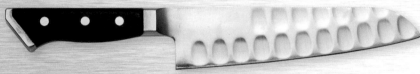

Santoku knife

(Also known as an oriental cook's knife or Japanese cook's knife)
A hybrid of a Western cook's knife and a Chinese cleaver (see page 17), the Santoku knife has a shorter blade but is wider than most cook's knives. These knives normally have a hollowed edge (as shown above) but occasionally have a straight edge. Beautifully balanced, making them easier to control, Santoku knives are great for sashimi (see page 148) and indeed all chopping needs, especially super-thin slicing, hammer-chopping (see page 30) and the fine, precise chopping of chillies.

Kitchen knife

(Also known as an office knife or fluting knife)
This small, popular knife is great for intricate jobs, and peeling, mincing and dicing in particular. Use instead of a cook's knife for finely chopping shallots. This knife is no good for the preparation of large ingredients.

Off-set handles

Some serrated knives have a step down in the handle. This is known as an off-set handle and prevents hands and knuckles being bruised when slicing.

Small serrated knife

(Also known as a tomato knife, fruit knife, deli knife or brunch knife)
Great for a variety of cutting tasks and foods such as crusty rolls, bagels, tomatoes and fruit, the tip of this knife is either rounded, pointed or has a forked shaped tip to assist with preparation.

Bread knife
Bread knives have rigid serrated or scalloped edges, or a combination of the two. The indentations allow the knife to cut through a hard crust without damaging the softer centre.

Carving knives
Carving knives go where straight-edged knives can't.

The blades are always long as the motion needed is one of sawing back and forth. The blade edges are also rarely straight but, as pictured here, serrated, scalloped or with a hollow edge; these edges are best equipped to deal with both tough and delicate ingredients.

Carving knife
(Also known as a scalloped slicing knife, ham knife or roast beef slicer)
The carving knife can be serrated, hollowed or straight edged but is almost always long and thin – perfect for making long, sweeping carving actions.

Salmon slicer
These knives usually have a hollowed edge to create air pockets between the oily salmon and the knife (so they don't stick to each other). With the long, thin and slightly flexible blade, they are perfectly designed for cutting thin slices.

Patisserie knife
(Also known as a pastry knife, larding knife or confectioner's knife)
A carving knife with a scalloped edge, the patisserie knife is also a specialist knife. The shape of the blade is similar to that of a palette knife and is perfect for cutting delicate pastries.

Specialist knives

The knives pictured here are for specific tasks and are probably not worth investing in unless you are very keen or a professional (see page 17 for *The Essential Eight*). Most tasks can be undertaken with a cook's knife, but specialist knives will maximize the potential of the knife, giving the best results, and add a professional touch.

Boning knife

This knife is mostly used by butchers (although it differs to a butcher's knife, see page 17). It's a rigid knife with a long thin blade that gets wider nearer the handle. The knife is designed to be used in a dagger-like motion (see page 35) and is perfect for boning meat and removing sinew and fat.

Turning knife

(Also known as a paring knife, peeling knife, shaping knife or bird's beak)

The curved blade of this knife is ideal for turning vegetables and shaping ingredients that have a curved surface. It is also useful for peeling an array of fruit and vegetables – a handy knife to have around.

Flexible knife

(Also known as a fish-filleting knife or a fillet knife)

This knife has a thin flexible blade, to fit around curved fish fillets and bones. The shape of the blade seems to vary between different manufacturers; it may sometimes look like a meat boning knife.

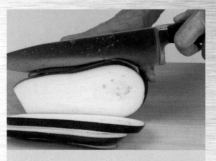

Left-handed chefs

Generally speaking, most knives are ambidextrous. However, some Japanese knives and serrated knives have edges on one side only, and so will not work for a left-handed person. Check before you buy – most manufacturers make knives for left- and right-handed chefs.

Meat cleaver

The large, weighty blade of this knife makes it very useful for heavy jobs and butchery. It is one of the few tools that should be used to chop through bones – if you use a regular knife it will damage the blade. A bone splitter, which is a cross between the meat cleaver and cook's knife, is also designed for this purpose.

Chinese cleaver

(Also known as a Chinese chef's knife or Chinese chopper)
Despite its intimidating appearance, the large, square-ended Chinese cleaver is surprisingly comfortable to use and offers precise and methodical chopping for a range of ingredients.

Palette knife

The blunt but flexible blade of this knife is great for lifting delicate pastries and slices of sticky cake, and manoeuvring around tricky situations.

Butcher's knife

The blade of this knife makes it a multipurpose tool for butchers. Some are rounded and plump at the tip and others are pointed.

Sashimi knife

Longer and thinner than Western knives with pointy tips, these knives are generally beveled on one side only to provide the sharpest edge. These knives are more expensive as they are generally handmade. They are also held differently, almost as an extension of your arm (see page 35).

The essential eight

There is a perfect knife or tool for every cutting task, although sometimes what we need may not necessarily be available or accessible. Below is our list of kitchen must-haves: The essential eight.

1 Cook's knife (p. 14)
2 Steel for sharpening (p. 25)
3 Kitchen knife (p. 14)
4 Turning knife (p. 16)
5 Bread knife (p. 15)
6 Boning knife (p. 16)
7 Flexible knife (p. 16)
8 Carving knife (p. 15)

Other cutting implements

Whether slicing soft cheese or dividing a pizza, carving carpaccio, creating julienned vegetables or shaving a truffle, there is a cutting tool to aid the task. Below is an overview of the most common kitchen utensils, that, along with your knives, will make culinary preparation of a variety of foods as quick and easy as possible.

Zester
A zester can take this form (left), but is available in a number of different shapes and sizes. Whatever their form, zesters are all designed for the same function: to remove the peel from the pith of the citrus fruit and turn it into fine zest.

Cheese knife and cheese slicers
There is a range of knives designed to slice the huge variety of cheeses available easily and neatly. An extra-strong cheese cutter, with a short and wide blade, slices easily through firm or semi-soft cheeses, including Gouda, Pecorino, Sardo and Cheddar. A long, fine-bladed knife works with ease to stab soft cheese with a washed or natural mouldy rind. A softer cheese will transfer more easily from a knife with a perforated edge than that with a smooth one. Depending on your choice of cheese, there is a knife to suit.

Cannelle knife
(Also known as citrus zester or orange peeler)
This tool is great for grating citrus zest and making stripy fruit and vegetables. The curved metal end is perforated with a series of round holes. By pressing the tip against the fruit with moderate force and drawing against the skin, the outer zest is separated from the pith beneath, and drawn through each of the holes into ribbons.

Melon baller (Also known as Parisian scoop)
Perfect for scooping and sculpting round balls, this tool hollows out the flesh from a variety of fruit and vegetables to create sphere-shaped portions. It works well with potato, melon, pear or avocado, and is great for deseeding a pear (see page 105).

Food processor
An extremely versatile appliance, the food processor makes life very easy. The majority have attachments to slice, dice, julienne, blend and juice. They are also available with built-in scales and are a great tool for making perfect breadcrumbs.

Poultry shears
A good strong pair of poultry shears (or indeed kitchen scissors) is a very useful addition to your kitchen tool kit. Great for cutting through fish and poultry bones, the blades of the scissors should be sharpened regularly (see pages 24–29).

Mezzaluna
The name mezzaluna (meaning 'half moon' in Italian) is representative of the tool's curve-shaped blade. The tool is most commonly double-bladed, with a handle at each end. It is rocked in a back-and-forth motion and used to finely chop herbs or pesto. Single-bladed versions are also available and may be used to cut pizza. The mezzaluna is also known as an hachoir, from the French verb 'hacher' (to mince).

Pizza cutter
An invaluable tool for dividing pizza into slices, the rotary blade of the pizza cutter (if sharp) is perfect for cutting through pizza bases without dragging the topping in the process.

Grater/microplanes
Fantastic for finely preparing a range of fruit and vegetables, as well as cheese, these tools also enable you to obtain fine zest. They usually have perforations of various shapes and sizes for different results.

Utility knife
The blade of this knife can be single- or double-ended and is retractable and adjustable – it can be extended to suit the task. This is the only tool suitable to simply score a pig's skin for roasting. If you don't have one, ask your butcher to do the job for you.

Mandoline
A kitchen utensil with blades of various shapes and sizes mounted on a fixed surface, the mandoline enables julienne of several widths and is great for creating wafer-thin vegetables. The only utensil to provide a uniform result each time, the mandoline is perfect for preparing foods that are deep-fried or baked, such as potato chips. Be sure to watch your fingers at all times.

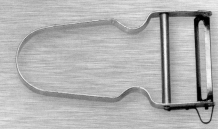

Truffle cutter
Similar to the above but on a smaller scale, the truffle cutter is designed to effectively shave paper-thin truffle slices, for pasta dishes or salads or to be inserted into meats or under poultry skin. As with the mandoline, take care with your fingers.

Peelers
An indispensable kitchen tool, the peeler is obviously designed for easy and economical peeling, but can also be used to make great vegetable 'tagliatelle'. Steel peelers are the easiest and most effective to use, but ceramic peelers are now available that will retain their sharpness for longer.

Meat slicer
Amazingly accurate for slicing thin carpaccio, Parma ham or sausages for melt-in-the-mouth, smooth textures, the meat slicer is a wonderful tool for obtaining thin, meticulously even slices.

Oyster knife
This specialist knife is the safest and most effective way to open an oyster's shell. For more on the methods involved in oyster preparation, see pages 154–155.

Fish tweezers
Intended to aid the removal of small or fine bones from fish, a pair of ordinary, clean cosmetic tweezers will essentially do the same job.

Fish descaler
Although not really necessary, this tool removes the scales from a fish. Some fish knives have this built in, although you can use the back of a standard knife to do the same job – see page 142.

Other useful items

As well as the huge variety of knives and array of other kitchen equipment available, there are a few additional items that a working kitchen should never be without.

Plastic gloves
A pair of plastic gloves is a very useful and hygienic addition. Keep on hand for messy jobs such as meat, poultry or fish preparation.

Kitchen towel
Always good for a quick clean-up of knives, chopping boards or kitchen surfaces, keep within easy reach. You can dampen slightly and place underneath your chopping board to stabilize too.

Salt
Roughly ground sea salt has great abrasive properties and is a useful aid to the crushing of garlic. It can also be used to provide additional grip – when skinning fish, for example.

Tongs
Great for gripping and lifting food, tongs are a useful kitchen implement to have at hand. They are ideal for use when barbecuing or frying food, or simply as a means of keeping hands and wrists out of the way of heat and spitting oil or water.

Chopping boards

A good chopping board is indispensable in the preparation of food and essential if you are to develop good knife skills. A sturdy counter may seem suitable as a cutting surface, but it will become damaged over time, and the lack of resilience will mean your knives also become damaged. For this reason, always use a chopping board.

There is an ever-increasing range of boards on the market – from those simply made of wood or synthetic materials to those designed to grip meat and catch juices, and boards that even have built-in drawers to store trimmings. Whatever the material, the ideal chopping board should be perfectly flat, solid and steady. A board that is worn (i.e. not flat) will make it difficult to chop evenly and this will inevitably affect your knife skills.

Choosing your board
Your chopping board requirements will depend largely on your diet and lifestyle. The best chopping boards are made from thick wood – the resilience of wood makes it an excellent material from which to cut on – and they will last a lifetime if they are well cared for. In an ideal world, you should have a separate chopping board for different food groups, although if space is minimal, you might want to just keep two boards – one for strong-smelling foods such as onion and garlic, and one for those with more delicate flavours. You should always be aware of cross-contamination between foods: strong-smelling foods are likely to leave a scent that may hang around and then later be transferred to other foods. Isolating these foods from those with a more delicate flavour will prevent such transmission.

Following is some useful information about the different boards available.

Wooden boards
Wood is the traditional material for chopping boards – it's sturdy, durable and its natural resilience means it is kind to the cutting edge of your knife. Wooden boards are available in end grain and flat grain wood. End grain boards are cut across the grain and as the blade of the knife is drawn across the board's surface, the individual wood fibres are able to bend out of the way, preventing them from damage. Because the knife is not cutting across the fibres the board will resist wear for longer.

Making your surface stable
For safe and effective chopping, always ensure your chopping board is perfectly stable. Some boards have grips attached to the bottom, but if your board is without, you can easily stabilize your surface by laying a thin, damp cloth or kitchen towel underneath the board.

Protect your knives
Whichever design you choose, bear in mind the ideal material to cut on is high-density plastic or any wood, from maple to beech. Never cut on china, marble, glass, stone, metal or kitchen tiles. These materials are too hard – the lack of resilience will damage your knife.

Wooden boards come in all shapes, sizes and a range of wood materials. Some are designed with grooves to catch juices and some with grips on the bottom.

Coloured boards are ideal to keep on top of hygiene matters and prepare food safely.

As the name suggests, the grain of flat grain boards lies flat. The simple construction of such boards means they are good value for money, but not as durable as end grain kinds.

Wooden boards do raise some questions regarding hygiene. There are some concerns that the porosity of the wood may promote the growth of bacteria, although it can be argued that the natural oils that protect the tree when it is alive still do their job as part of the board. This aside, there is no reason why properly sanitized wood should not be hygienic. With any board, you should always be aware of bacteria and contamination – cleanliness should be at the forefront of your mind during all food preparation.

Synthetic boards

Synthetic boards are generally inexpensive, and much easier to clean and care for than wooden boards. They are not as kind to knives as wooden boards, but because they are lighter and thinner they are much easier to store. A good quality board should also prove long lasting. Note that poorer quality, thin plastic boards may become worn over time, especially if put through the dishwasher, and as a result may become unstable to work on.

Colour-coded boards

In many professional kitchens, colour-coded synthetic boards are used. Each colour is designed to be used for a certain food type only, i.e. blue for

fish, red for meat, green for fruit and vegetables. This colour-coded system is a great step towards food safety and makes it easier to ensure cross-contamination is avoided; however, there is no substitution for thoroughly cleaning your chopping board between uses, so clean your board each time.

Cleaning your chopping board

Whatever the material of your chosen chopping board, it should be kept perfectly clean. As with knives, the best way to clean your board is to hand wash it in warm water with a mild detergent, before patting dry. Wooden boards should never be left to soak or be put through the dishwasher – the wood will crack if it becomes too hot. Some synthetic boards may be dishwasher safe, although no board should really be subjected to such harsh temperatures – they may cause warping. When cleaning a board that has been used in the preparation of meat or fish, bear in mind that hot water can 'cook' any food remnants, almost gluing them to the board. To prevent this, rinse the board under cold water first to remove any residue, before scrubbing with hot soapy water as usual.

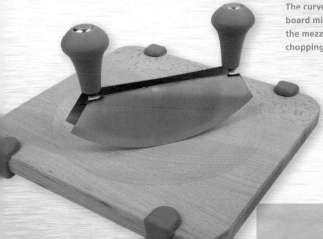

The curved surface of the mezzaluna board mirrors the curved blade of the mezzaluna perfectly, making the chopping of ingredients harmonious.

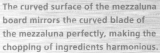

Sharpening knives

There is a distinct difference between the results of a sharp knife and those of a blunt one. A sharp knife will effortlessly glide, producing well-defined cuts and ensuring a professional and uniform end result. More pressure has to be used with a blunt knife in order to gain the same level of performance. The increased pressure makes the knife more likely to slip, making your knife-work sloppy and less precise, as well as dangerous. To get the most out of your knife, ensure it is always sharp.

Tip

For the sharpest possible knife, do as the Japanese do. Every night after service, they sharpen their knives on a whetstone, then oil and polish them, almost ritualistically. They rarely (if ever), use a sharpening steel.

Maintain a constant angle

Whichever sharpening method you choose, the most important thing is maintaining a constant and accurate angle of between 12° and 20°. Some whetstones provide an angle guide to assist this.

The knife edge: what makes it so sharp?

Consider the edge of a piece of paper and how incredibly sharp this can be. This is because it is so thin. The same applies to the blade of a knife, but the thin edge will be blunted a little every time it is used, making it thicker. For optimum performance, the edge of the blade must constantly be redefined in order to maintain its sharpness. It is hard work to keep a knife sharp, but chefs with the best knife skills are undoubtedly those who take time and care over keeping their knives razor sharp. There is no getting around it: a brand new knife will be perfectly sharp but through regular use the edge will start to blunt quickly.

How often should a knife be sharpened?

The simple answer is as often as is needed. To illustrate this point, take some onions and a brand new knife. Start to chop the onions and by the end of the second onion, or thereabouts, the sharpness of the blade will start to disappear as the edge begins to blunt. This will give you an idea of how necessary it is to sharpen knives, especially in the professional kitchen. In a domestic environment, get into the habit of sharpening your knife every time you use it. The majority of knives are made from steel, and need to be sharpened regularly. Ceramic knives keep their edge for longer and don't need to sharpened as regularly. Always ask the manufacturer or the shop assistant when you buy the knife – they will recommend the best method of sharpening and advise on the regularity.

Don't lose the edge

To get the most out of your knife, it is essential to sharpen it regularly. After a period of time, a knife that hasn't been sharpened will lose its edge. The only way to get this back is by taking it to be sharpened professionally or by using a whetstone (see page 26).

Sharpening tools

Old school barbers would often sharpen their blades on a strip of leather. Knife-sharpening methods today include using a steel (ceramic, diamond-coated or traditional ridged/honing), a sharpening wheel (waterwheel, diamond-coated or ceramic) or a whetstone.

Steels

Steels are the most widely used sharpening tool. They offer a quick fix to the blunted edge of a knife and have the advantage of being portable.

Diamond-coated steel

Because of its abrasiveness, this is a very effective sharpening steel, however, in the wrong hands, it can wear down the exterior of the knife and take years off its life, so be sure to find out how to utilize this tool correctly. The diamond coating will become worn over time, and so like traditional steels, it will need replacing. This type of steel offers a good quick fix, but in the long-run, maintaining sharpness with a honing steel or a whetstone is ideal.

Ceramic steel

The ceramic steel combines the best of both the honing and diamond-coated steels. It offers ridges like the traditional honing steel, but also a fine, abrasive surface like the diamond-coated steel. This steel is considered the most superior kind.

Traditional ridged or honing steel

The most common type of steel to come in a knife set, this steel is effective with most knives. It is essential to use this steel regularly if you are to keep a good sharpness on the blade. If you don't use regularly, the edge of your knife will be lost and no amount of using this steel will bring it back: the knife will become useless. If this does happen, consider using the whetstone or taking to a professional to rectify. These steels will wear over time, and need replacing.

USING THE STEEL METHOD

Step 1
Grip the steel in one hand comfortably. With the other hand, hold the knife and rest the heel of the blade at the bottom of the steel.

Step 2
Maintaining an angle of between 12° and 20°, pull the knife gently backwards and upwards along the steel. When you reach the middle of the blade, the knife should be in the middle of the steel.

Step 3
Continue along the steel until you reach the end, and the tip of the knife.

The professionals

As an alternative to doing the job yourself, you can get your knives sharpened professionally. Be aware of the quality of these services before you use them – sometimes too much of the edge can be taken off a knife. Kitchen shops offer a sharpening service, as do some butchers – they tend to have their own grinding wheel.

USING THE WHETSTONE METHOD

The whetstone is without doubt the best sharpening tool. Although the skills for using this tool take more time to master, it does ultimately give the best results, providing it is used correctly. To use a whetstone effectively is really quite an art. Made from either ceramic or stone, whetstones are heavy and also bulky to travel with. Some need to be soaked prior to use or remain under a constant flow of water during sharpening. When using this method of sharpening, it is important to maintain a constant angle between the knife and the whetstone. The whetstone pictured here has a clip that you can attach to the knife to maintain a constant angle.

Step 1
If applicable, soak the whetstone in water until the bubbles disappear (for approximately 1 hour). The whetstone may be contained in a stand (as pictured in step 2) which will steady it. If it isn't, place the stone on an old towel to steady it. Make sure it is protruding away from you.

Step 4

Now begin the process again from step 1, but on the other side of the knife. Continue for as long as is needed to sharpen the blade.

Honing the knife using a surface as a support

For those with less experience, rest the steel down on a surface and hone your knife in exactly the same way as shown opposite.

Always check the angle of the blade. An angle greater than 20° can damage the blade.

SHARPENING WHEELS

Step 2

Start with the heel of your knife at the far end of the stone. At an angle of between 12° and 20° to the whetstone, drag the knife in a backwards direction (towards you), moving the knife across the stone at the same time. You don't have to apply too much pressure, but remember to keep the angle constant.

Step 3

Continue this process on one side, until you feel a rough edge on the other side. At this point, turn the knife over and do the same to the other side. Do this at least four times to each side, then test the sharpness (see page 29). If it is not sufficient – simply continue. Clean the knife and the whetstone afterwards.

Sharpening wheels are basically two steels that are set to a rigid angle on both sides. They will need replacing after a while. They can be diamond-coated or ceramic and sometimes use water. The sharpening wheel is not the most respected or widely used sharpening tool, but it is guaranteed to give good results if used properly.

Ten rules for sharpening knives

1 Choose your method

Decide if you want to hold the steel in your hand outwards or downwards, using your work surface as a support. (The latter is the easiest way to maintain a constant angle.) If using a whetstone or a sharpening wheel, find a suitable working space.

2 Right your posture

Stand with your feet hip-width apart and bend your working arm at the elbow.

3 Relax

If using the steel method, grip the steel firmly but comfortably in one hand. There is no need to put too much pressure on the steel or the knife – a moderate contact is all that is required. The same applies when using the whetstone or sharpening wheel – relax and work comfortably.

4 Be hygienic

Always sharpen your knife in an isolated area – filings can spray over food.

5 Check your angle

The angle of the knife is very important: it should be small, ideally between 12° and 20°. If your angle is too big, the blade will be damaged. If possible, check this with the manufacturer when you buy the knife.

Serrated knives

There has always been a debate as to whether you can sharpen a serrated knife or not. There is a school of thought that if only one side of the knife is serrated, then it is possible to sharpen. There are some new sharpeners on the market that claim to sharpen serrated knives, although most manufacturers maintain that you cannot and should not sharpen a serrated knife. A good domestic serrated knife will last for years.

6 Sharpen the whole knife

Sharpen the full length of the blade. Start with the heel of the knife and gently pull backwards while moving up and across the knife.

7 Sharpen both sides

Once you have sharpened one side, begin again on the other. It's important to apply exactly the same process to both sides in order to keep the knife's edge even.
NOTE: If the knife is bevelled on one side, sharpen the bevelled side only.

8 Test it

To check if the knife is sharp enough, try slicing something, be it a tomato, potato, onion or apple. The knife should glide through easily and meet no resistance. If it doesn't, simply continue with the sharpening process until it does.

9 Be patient

It is difficult to estimate how many times you need to repeat the sharpening process – do so until the knife is sharp enough (see rule 8).

10 Clean up

Remember to clean both the knife, the sharpening tool and the surrounding area when you have finished sharpening.

Core cutting methods

In this section you will find step-by-step instructions on how to properly execute key methods of chopping that are mentioned throughout the book. Using the appropriate cutting method (and grip) in the correct way will ensure that you can prepare all kinds of food efficiently, precisely and safely.

Chopping rules

- Always use a sharp knife.
- Use the right knife for the right job: this maximizes the knife's potential.
- Position the board at a height suitable for you – check your posture and make sure your back is straight (see page 35).
- Ensure the board is steady – put a damp cloth or kitchen towel under the board (see page 22).
- Whatever you are chopping, always try to lean it on its flat surface. Slice off an edge to gain one, if necessary.

HAMMER-STYLE CHOP

The hammer-style chop is difficult to develop – it takes time, experience, precision and a very sharp knife. For best results, implement this chopping method with a cook's knife, a Japanese Santoku knife or a Chinese cleaver (as shown to the right). The shorter and fatter the blade, the easier it is to balance. Needless to say, always tuck your fingers well out of the way. Although demonstrated here with a butternut squash, you should practise this cutting method on a soft ingredient, such as a white mushroom. As with all things, practice makes perfect.

UNDER-THE-BRIDGE CUT

If conducted properly, this is probably the safest cut, as fingers that are used to grip the ingredient are kept well out of the way. This is a great method for cutting potatoes. A kitchen knife or any cook's knife works well with this cutting method.

STEP 1

Use the thumb and the first finger to grip the knife handle, and use the three remaining fingers as a support to make the cut. Use the claw grip to steady the ingredient (see page 32), before cutting down.

STEP 2

Work the blade up and down, and back and forth to chop.

STEP 3

Like the rolling chop (see page 32), the blade should just skim the knuckles of your fingers as it goes up and down.

STEP 1

Grip the ingredient with the fingertips and thumb of one hand, and raise the palm of the hand to create an arch, or bridge. Insert the knife under the bridge and between your fingers.

STEP 2

Place the tip of the knife on the board and cut downwards as shown.

STEP 3

Still using your fingers to grip, apply sufficient pressure to cut right down to the chopping board, making a clean cut.

ROLLING CHOP

The rolling chop is one of the most common cutting methods, and is most notably used in conjunction with the claw grip (see below). The knife tip is used as a support and constantly moves, but maintains contact with the board throughout. The whole length of the knife is brought backwards and the cutting process starts only once the knife is moving forwards in a rolling motion. In most cases, the ingredient gets the full benefit of the knife's blade. A long knife, such as a cook's knife, works particularly well here.

STEP 1
Steadying your ingredient using the claw grip, start with the tip of the knife on the chopping board. Pull the knife back and down.

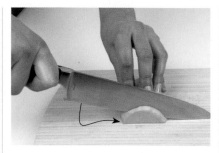

STEP 2
When the knife reaches the ingredient, change the cutting direction of the blade, and using the belly of the knife, start to cut forwards, while cutting through the ingredient.

THE CLAW GRIP

Probably the most common method of gripping an ingredient, the claw grip is so called because the hand is tucked into a position that resembles that of a claw. The little finger and thumb are used to grip the ingredient.

STEP 1
Practise forming your hand into a claw – it's important to feel comfortable before implementing the grip.

STEP 2
Now grip the ingredient using this method. The aim is to position the hand so that when the knife is used, it just brushes the knuckles (see step 2 of *Hammer-style chop*, page 31).

STEP 3
When the knife reaches the flat of the board, start to pull the knife upwards.

STEP 4
When the knife is fully at the top, start to pull back, again with the tip of the knife on the board.

STEP 5
Repeat the whole rolling motion again.

LONG SWEEPING SLICE
The long sweeping slice involves making sweeping movements to create thin, elegant slices. Perfect for slicing smoked salmon, it is essential to use a long, thin knife with a hollowed or fluted edge to make this cut.

STEP 1
Start at the heel of the knife and at one end of the ingredient. After inserting the knife, pull the handle backwards slowly.

STEP 2
When the tip of the knife is reached, simply change direction and go the other way (forwards). Continue making your way along the ingredient to produce a thin, precise slice.

LEVER CHOP

The lever chop involves inserting the knife at a 90° angle to the ingredient and using the whole length of the knife to cut straight down. Using a knife in such a way is not generally seen as good practice, although in some circumstances this is the only way to correctly prepare certain ingredients. For best results, try to use a knife with a length that exceeds that of the ingredient.

STEP 1
Start with the knife at a 90° angle to the ingredient, so you are ready to move in the direction marked.

STEP 2
Steady the ingredient with your free hand, and using the whole length of the knife, cut downwards. As you cut your way through the length of the ingredient, the angle between the knife and the board should steadily decrease.

CROSS CHOP

Similar in style to the cuts made by a mezzaluna (see page 19), this quick cutting method is perfect for chopping herbs and mincing meat. One hand grips the handle of the knife and the other steadies the tip, which stays in one place while the handle is moved around and pulled up and down quickly. Once the ingredient is chopped, it is scraped up again with the blade and distributed over the board again, so the process can be repeated to refine the chopped pieces. A large cook's knife is ideal here. You can of course use a mezzaluna, if you have one.

STEP 1
Using the claw method to grip (see page 32), start by roughly chopping the ingredient until the pieces are of a manageable size.

STEP 2
Now place the heel of one hand over the tip of the knife (to steady it) and grip the handle in the usual way. Keeping the tip of the knife in one place, move the handle from left to right and continue to chop, mainly using the heel of the knife. Stop intermittently and redistribute the ingredients all over the board again.

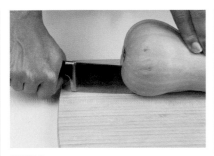

STEP 3

Continue to cut in this downwards direction, until the knife is flat on the chopping board.

STEP 1

It is important that you maintain good posture during chopping. Position the chopping board at the correct height so that you can comfortably stand up straight when chopping The board should be around waist height.

STEP 2

During chopping, relax your wrists and shoulders. Work with as much clear space around you as possible. It must be comfortable for your knife to move if it is to work to its full advantage.

STEP 3

Carry on chopping like this – from left to right and then right to left – until your ingredient pieces are of the desired size.

THE BONING KNIFE

Although the boning knife can be held and used like a regular knife, it is designed to be used in a dagger-like way, as shown above. This enables the sharp curved tip to be used to its full advantage. The rigid blade allows precise cuts to be made.

THE SASHIMI KNIFE

The sashimi knife should be held with the handle tucked well against the wrist. When using, position the first finger along the top of the blade to provide support. This knife should then be used as though it is an extension of your arm.

vegetables

and herbs

From the simple onion to the more complex artichoke, a wide range of knife skills is required for the variety of vegetables and herbs available today. Whether mastering techniques like the paddle method of preparing garlic, or the ability to craft the perfect carrot baton, this chapter will demonstrate all you need to know about the preparation of these foods.

Vegetable and herb knives

As vegetables and herbs are a range of different shapes, sizes and consistencies, so are the knives designed to prepare them. Start with a knife that feels comfortable to use and which suits the food you are preparing.

Kitchen knife

A small, all-purpose knife, this is basically a smaller version of the cook's knife. With its blade measuring approximately 5 cm (2.5 in.), its small size enables greater control, meaning it's the ideal knife to use as you begin to learn new skills. Perfect for learning to finely chop a shallot, start by using this unintimidating knife for all sorts of tasks before graduating to a larger cook's knife.

Other useful tools

MUSHROOM BRUSH AND KNIFE
For cleaning and peeling mushrooms.

MANDOLINE
For creating wafer-thin vegetables and uniform julienne of various widths (see page 20).

CANNELLE KNIFE
For making stripy vegetables (see page 18).

MEZZALUNA
Great for roughly chopping herbs and sometimes garlic, the mezzaluna is good fun to use and gives informal results that are perfect for rustic food at home. It can be single- or double-bladed, and with a handle at each end is used in a back-and-forth cutting motion (see page 19)

Chinese cleaver

The sheer quantity of beautifully chopped vegetables in Chinese cuisine reveals just how perfect this tool is for chopping vegetables. It has a sturdy handle and a huge and heavy rectangular blade, but don't be put off by its size: this versatile knife is great for chopping vegetables, as well as poultry and mincing meat.

Cook's knife/Santoku knife

The most commonly used type of knife, the traditional cook's knife (with a 15–30 cm [6–12 in.] long blade) is a great all-rounder for preparing a wide variety of vegetables and herbs. A Santoku knife (see page 14) may be preferable for more precise chopping. Choose the size most suitable to you.

Turning knife

Originally developed for using when turning vegetables for classical French cuisine, this knife is a great aid to the basic knife skill of turning (see pages 74–75). As well as making turned vegetables (which are easy to cook and present), the curved blade of this knife is equally useful for all sorts of intricate peeling and shaping jobs.

Small serrated knife

The perforated edge of this knife copes beautifully with tough outer skins, making it the perfect knife to use with vegetables such as tomatoes, aubergines, chillies and peppers. The serrations of the knife work equally well on delicate soft-fleshed fruits such as plums, peaches and nectarines.

Vegetable peeler

Fantastically useful, a vegetable peeler not only makes peeling easy, effective and less wasteful, but also allows you to turn vegetables into vegetable 'tagliatelle' or to thinly slice them for 'fritti' (see page 55). Today, ceramic peelers with an even sharper blade are available.

Onions

Onions serve to flavour a variety of foods, and as one of the most widely used vegetables in cooking, learning how to cut an onion is an essential knife skill.

To get the best quality out of your onions, they should be cut just before you are going to use them. Try to cut the onions in a uniform manner and avoid 'cross chopping' the onion (see page 34). This produces inconsistent results and increases the gaseous aroma.

WHICH KNIFE?
Slicing and chopping

Use a medium-sized cook's knife for both of these tasks. For larger varieties, such as Spanish onions, you may want to use a large cook's knife. For shallots, use a small cook's knife or kitchen knife.

Avoid watering eyes
The gaseous substance *allyl sulphide* is released when an onion is chopped, and it is this chemical that makes the eyes water. Many cooks have their own tested method of avoiding these tears, including wearing goggles. Try freezing the onion for 10 minutes before preparation. When the onion is cold, little aroma is given off, thus reducing its effect.

Chopping shallots

Similar in shape, the shallot differs slightly from the onion because of the interior bulbs that make up the vegetable.

Peel, chop and handle shallots in the same way as onions, keeping the root intact during preparation.

SLICING

Step 1
Grip the onion in one hand and steady it. Cut off the papery end of the onion to produce a flat base.

DICING

Step 1
Take a peeled, halved onion with the root still intact (as in step 3 above). Make cuts lengthways through the onion, using the tip of a cook's knife. Take care not to cut right through the root end of the onion, but ensure that the knife cuts to the bottom of the onion (i.e. through to the board).

Step 2
Stand the onion up on its flat base, cut in half, straight through the root. Aim to go directly through the centre of the onion, to produce two equal halves.

Step 3
Peel each of the onion halves, trimming the root ends but leaving the root intact. This will make it easier to maintain a regular-sized slice. Place both halves flat-side down onto your chopping board.

Step 4
Secure the onion using the claw method (see page 32). Steady the tip of your knife on the board and use a rolling chop (see page 32), to slice through the onion. Make sure you use the full length of your knife, in a forward cutting motion.

Step 2
Continue to make cuts lengthways all the way across the width of the onion. As you begin to make your way across, change to the under-the-bridge method of cutting (see page 30). Bear in mind, the closer the cuts are together, the finer the end result. For larger dice, make the incisions further apart.

Step 3
Next, make three or so horizontal cuts into the onion as shown. Start from the bottom of the onion. The number of cuts you make depends on how small you want the dice to be. For larger dice use fewer cuts, spaced further apart.

Step 4
Now cut vertically down using the rolling chop (remember to use the entire length of your knife). Be sure to make definite cuts throughout the onion cutting process. You can cross chop at the end if you wish too, but this increases the unpleasant gas aroma from the onion.

Spring onions and leeks

The strong flavour of spring onion is disseminated wonderfully when chopped finely into rounds or sliced into thin julienne. Leeks are much milder in flavour, but the same knife skills can be applied.

There is a continuous debate about whether to use the green parts of these vegetables or not – some recipes may require you to, others may not. Whichever part you decide to use, the layers are likely to unravel during preparation. As with the other members of the onion family, keeping the root intact while you are cutting helps to maintain a regular size of chop.

Step 1
To slice or finely chop a spring onion, begin by trimming off the root ends. At this stage you may want to remove an outer layer of the onion.

SLICING LEEKS

Taking one leek at a time (as opposed to spring onions, which can be sliced in bunches), grip using the claw method and slice the leek into rounds of the desired thickness.

WHICH KNIFE?
General preparation

Use your favourite cook's knife or Santoku knife for preparing these vegetables. Use in conjunction with the claw grip and rolling chop. The Chinese cleaver is also good for assisting in very fine julienne.

Asian variation

For an authentic Asian preparation of spring onions, tilt your knife to an angle smaller than 90° and slice the spring onion at this angle.

FINELY CHOPPING A SPRING ONION

Step 2
Take your bunch of trimmed spring onions and grip using the claw method (see page 32). Start at the white end of the onion and using the rolling chop, slice the onion into rounds of the desired width. These rounds are great in a salad or scattered as a garnish.

Step 1
Start at the white end of a trimmed spring onion and, using the tip of the knife, pull backwards to slice the onion into thin lengths. If you need a very fine end result, slice closely at this stage.

Step 2
Rotate the spring onions 90° and slice with a rolling chop. Continue as far along the onion as you wish – you may or may not want to use the green part.

JULIENNE OF LEEKS

Step 1
Julienne is a popular method for deep-frying or stir-frying leeks, although this knife skill really tests precision. Begin by slicing the leek into 10 cm (4 in.) pieces. Halve each piece lengthways.

Step 2
Each piece will be made up of several layers. Unravel these as shown.

Step 3
Take a few layers at a time and pile them up, turning them over so the curved outside is facing upwards. Using the claw grip, slice the pile into julienne strips as thinly as desired.

Garlic

Garlic is most often used as a flavouring agent in a number of foods, but it can also be eaten as a vegetable. When used in its raw form, it is pungent and slightly bitter, but becomes much milder and sweet when sautéed or baked.

After the bulb has been separated, garlic cloves can be prepared and used in a number of different ways: they can be cooked whole, or alternatively sliced, chopped or crushed. The intensity of flavour very much depends upon the method of preparation. Chopping the garlic into smaller pieces, for example, will allow more of the juices to be released. This will consequently provide more flavour than slicing, and crushing the garlic will release more of the flavour still.

WHICH KNIFE?
Slicing, chopping, crushing

For more control, or for the less experienced, use a kitchen knife for slicing and finely chopping. Otherwise use a medium-sized cook's knife or a Santoku knife (as pictured in the sequence shown opposite). For crushing, a large cook's knife is essential to get plenty of movement into the preparation.

'Bruising' a clove

This is an ideal technique for roasting garlic, as keeping the skin on the clove protects it from burning. Instead of removing the base of the clove as in step 1 of Peeling (opposite), simply apply step 2 to an entire clove.

SEPARATING A WHOLE BULB

Place the whole bulb on your chopping board. Turn the garlic onto its side and place your hands on top of the garlic, one over the other as shown. Using your body weight, push down onto the garlic. The cloves should fall apart.

CHOPPING FINELY

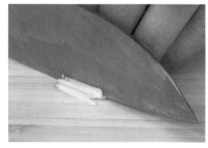

Step 1
Take a pile of sliced garlic and slice further into thin julienne.

PEELING A CLOVE

Step 1
Remove the base of the garlic clove by slicing the end off with a cook's knife.

Step 2
Cover the clove with the flat end of your cook's knife and, using your body weight, press down with the heel of your hand. This will open the clove up slightly so the papery skin can be easily removed.

SLICING

Holding the peeled clove carefully with the claw method (see page 32), slice through the garlic as thinly as you need. You can do this either lengthways or widthways.

Step 2
Turn the julienned pile 90° and slice again into fine petit brunoise (see *Glossary*, pages 172–173).

CRUSHING

Step 1
Roughly chop a peeled clove of garlic until it is in small pieces. Sprinkle with a pinch of salt. Move the garlic near to the edge of the board to allow more movement for step 2.

Step 2
With the heel of one hand at the tip of your knife and the other on the handle, press the thickset of the blade onto the garlic, and at the same time scrape it onto the chopping board. The motion resembles that of rowing a boat: this is known as 'the paddle method'. Continue until the garlic forms a smooth paste on your board.

Carrots

Cooked, or used raw as a starter or in a salad, the carrot is an indispensable vegetable. With its sweet flavour, it is a vital and basic ingredient for stocks, sauces and casseroles.

Carrots can be sliced or chopped into the desired shape, or grated to add to salads. Whatever your preference for eating carrots, the preparation of these versatile vegetables will only ever involve basic procedures. That said, applying proper knife skills will ensure preparation is quick and effective. 'Blocking-off' a carrot simply entails removing the curved edges of a carrot to create a uniform rectangle shape. This ensures a precise and equal result for batons, fine julienne or dice, and so is a very useful technique to master.

WHICH KNIFE?
General preparation

For general preparation a cook's knife is ideal – choose the size you feel most comfortable with. A Santoku knife is great too. For more control a kitchen knife is useful and also a good substitute for a peeler.

If you are turning the carrot into a barrel shape use a turning knife (see *Turning Vegetables*, pages 74–75).

Keep the goodness

Many vitamins are contained in carrot skin and so you may not always want to peel them. Instead, preserve the goodness by scraping with a knife or a scourer to clean the skin.

Carrot batons

Batons are difficult to perfect by hand but are much tastier (and fresher) than commercial versions, and are great for crudités and dips.

PEELING

The easiest way of peeling a carrot is to use a vegetable peeler, as shown above. To remove the skin using a knife, hold the carrot at the tail end and lean the top onto your cutting board. Using long sweeping movements, and working away from your body, turn the carrot until all sides are peeled.

CHINESE STYLE

Step 1
This is an attractive way of preparing carrots. Cut the tail end off a peeled or scraped carrot. Turn the carrot 90° and cut so the resulting carrot piece is triangle-shaped in profile. If particularly thick, cut in half lengthways first.

SLICING

Step 1
Using the claw method (see page 32), grip the carrot between your fingers, using the knuckles of your middle fingers as a support for the knife. Slice the carrot into rounds of the desired width, using a rolling chop (see page 32).

Step 2
You can also execute this method on an angle. This is known as cutting 'on the cross' or 'oblique'. When cutting on the cross, use the same angle as shown in Chinese style, step 1.

Step 2
Each time you create a triangular piece, roll the carrot over and cut again. Continue with the process for perfect Chinese-style carrots.

CARROT BATONS

Step 1
Take a blocked-off piece of carrot and cut lengthways into a flat slab measuring 1 cm (½ in.) in width.

BLOCKING-OFF

Cut the carrot into lengths (the combined width of four fingers, approximately 8 cm [3 in.] in size). Take each piece, stand it on one of its flat ends and block-off by slicing away the curved side. Repeat until you have four straight sides. After the first curved edge is removed, lay the carrot on its longer side.

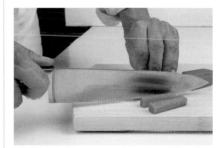

Step 2
Turn the slab over so its biggest side lies flat on the chopping board. Cut further into batons, 1 cm (½ in.) in width. For a perfect baton you need to be accurate, so don't rush! Batons should be a uniform rectangle shape (approximately 1 cm x 1 cm x 8 cm [½ in. x ½ in. x 3 in.]).

Common brassicas

The shapes of broccoli and cauliflower vary according to the variety, but generally these vegetables should have tightly packed, dense heads and feel firm to the touch.

In the first stages of preparing a brassica, it is best to use a kitchen knife or a small cook's knife, as the tip of the knife can be used to easily deal with the florets that are tightly packed against the stem. The real skill of preparing broccoli or cauliflower lies in maintaining the natural floret shape as much as possible. Try taking the time to shape the ends of floret stalks into pointy tips.

BROCCOLI

Step 1
Remove the outermost florets first and rest the head of the broccoli on the chopping board. Hold the thick stalk in one hand. Using a kitchen knife or small cook's knife, slice away from you at a 45° angle to the stalk, towards where the floret joins the main stem. Put to one side to trim later.

CAULIFLOWER

Step 1
Hold the cauliflower in one hand and, with the other hand, snap off the leaves at the base.

WHICH KNIFE?
Separating and trimming

For the initial separation of florets, a kitchen knife or small cook's knife is ideal. Continue to use this for trimming the florets too.

For purple sprouting broccoli, use a medium cook's knife to trim the stems. When preparing the stalk of broccoli, use a large cook's knife or a Santoku knife.

Purple sprouting broccoli

As purple sprouting broccoli grows, the stem becomes thicker and 'woodier'. Trim to leave even and tender stems – generally up to about one-third should be removed.

Step 2
Turn the broccoli and continue removing the florets. They will become less defined the nearer you get to the top of the head.

Step 3
Once at the end of the stem, slice the remaining top floret off. Divide this large piece further if necessary. (You may find it easier to first cut through the stem and pull the head apart with your fingers.)

Step 4
Sometimes the florets are naturally larger than bite-sized. If this is so, cut through the stem first and divide the floret further if necessary. Go back and trim the florets as desired. Any trimmings can be used for a stir-fry, alongside the stalk (see below).

Step 2
Gently turn the cauliflower on its head. With the point of a small cook's or kitchen knife, cut halfway into the stalk. Keep the knife in this position and carefully turn the cauliflower 360° to cut away the core.

Step 3
Remove the (now pointed) core and prepare in the same way as above. For a rustic approach, quarter the cauliflower into 'wedge' portions by standing it on its base and cutting down into four using a large cook's knife.

Using the stalk

The often-neglected broccoli stalk can also be used in recipes. Prepare the thick stalk by either peeling it with a vegetable peeler and then thinly slicing it, or by simply 'turning' it (see pages 74–75).

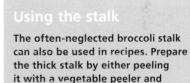

A hardy vegetable, the cultivated cabbage is available all year round. Unless very young, the core is unpleasant to eat, so all knife skills are concentrated on the leaves. White or red, because of their tightly packed leaves and size, a large cook's knife is required to get the sufficient height needed to slice these vegetables. The leaves of the savoy cabbage are less tightly packed and are consequently a little more forgiving to prepare.

Brussels sprouts are similar in formation to the cabbage, and are almost mini versions. Little in the way of preparation is needed, other than removing the base and outer leaves, but using a knife is still required.

WHICH KNIFE?

Cabbage

Use a sharp, large cook's knife to initially cut a cabbage in half. For further cutting, this can be substituted for a medium-sized cook's knife.

Brussels sprouts

To prepare a Brussels sprout, a kitchen knife is all that is needed. If, however, a rough chop of sprouts is required, upsize to a larger cook's knife or Santoku knife for easier and more effective handling.

Whole cabbage leaves

Cabbage leaves are great for wrapping around a variety of fillings, although only the outer leaves are suitable for this. Trim the cabbage at its base (as shown above), removing any leaves that fall away. Repeat until the leaves become compact.

CABBAGE

Step 1
Turn the cabbage on its side, steady with one hand and, tucking your fingers out of the way, trim the base using a large cook's knife.

BRUSSELS SPROUTS

Step 1
Rest the Brussels sprout on a chopping board. Using a small kitchen knife, slice the base off close to the root (being careful not to remove too much). Remove any tough outer leaves.

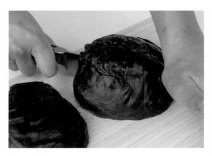

Step 2
Halve the cabbage into two pieces. Turn each of the halves onto their flat sides and halve again into quarters.

Step 3
Remove the core from each quarter by placing the knife just above the core and cutting on the diagonal.

Step 4
Place each quarter flat-side down (curved side facing up). Use the claw method and the rolling chop (see page 32) to slice evenly.

Step 2
If the Brussels sprouts you are using are very young (i.e. small), they simply need to be trimmed at the base. If they are larger and older then the outer leaves and a greater proportion of the base need to be removed.

Step 3
Be careful not to cut too far up the sprout, as all the leaves will fall off, leaving only a small amount of sprout. This will cause the leaves to overcook and turn a yellow-brown color before the main body is cooked.

Step 4
Make a cross incision from the edge of the base toward the centre, no more than 5 mm (⅕ in.) deep. This will ensure that the sprout cooks evenly and right through to its core.

Celery and fennel

Raw in salads or as a vital ingredient for a stock or sauce, these umbelliferous vegetables are 95 percent water, and as a result are best prepared at the last minute.

Both of these deliciously crunchy vegetables require assertive, definite knife skills. Raw fennel is tough and fibrous so is best enjoyed cut very thinly, either on a mandoline or with a cook's knife using practised knife skills. Bear in mind that fennel will discolour if exposed for any length of time; storing celery in cold water will help retain its crispiness, and prevent it from naturally curling.

WHICH KNIFE?
General preparation

For peeling 'stringy' celery a peeler is ideal, but you can substitute this for a turning knife or kitchen knife. For general chopping of both fennel and celery a medium-sized cook's knife or a Santoku knife works well.

DICING CELERY

Step 1
Trim the base of the celery and the leaves at the top. Take care not to remove too much of the body. Discard the stalk, or use in stocks or soups.

PEELING CELERY

To remove any stringy lengths, simply peel the celery with a vegetable peeler. Steady one end of the celery on the chopping board and peel away from you. You can also lay the celery flat on the board and peel lengthways.

SLICING FENNEL

Step 1
Lay the fennel on one of the longer sides, and steady in one hand using the claw method. Trim the base of the fennel. Try to remove as little as possible, but make sure the entire base is removed, and discard.

Step 2
Cut the stalks in half or into more manageable lengths.

Step 3
Slice each piece in half lengthways. Find a flat side and cut again either in half or thirds lengthways depending on the desired size of the dice.

At this point you can also use the celery as batons for crudités, dips, etc.

Step 4
Using the claw method (see page 32), group the batons together and slice them into small dice.

DICING FENNEL

Step 2
Start to pull the fennel 'leaves' away until the whole fennel is separated. You may have to trim the base even further as you go on, to enable the inner leaves to be pulled away. The inner core is dense and will have to be chopped separately.

Step 3
For slices, take each fennel leaf and lay curved-side down onto the chopping board. Using a rolling chop (see page 32), slice each leaf into strips approximately 1 cm (½ in.) in width (or however wide your recipe suggests).

For dice, follow steps 1–3 opposite. Then, group these strips together, turn 90° and cut into dice. For even squares, cut the same width as you did in step 3.

Courgettes

The soft but firm texture of courgettes, by comparison to its tougher squash cousins, makes it relatively easy to prepare.

Courgettes will quite happily be prepared ahead of time. Yellow or green, courgettes are usually up to 20 cm (8 in.) in length, with soft skin that can be easily pierced with a fingertip. Sometimes during preparation the moist flesh may stick itself to the knife, so a knife with a hollowed edge is the ideal knife to use here. A marrow is merely an older and larger courgette with thicker skin and more watery flesh, and can be prepared in much the same way as a courgette.

WHICH KNIFE?
Slicing and dicing courgettes

A hollowed-edge knife, if you have one, is ideal for the preparation of courgettes, as the flesh doesn't stick to this type of blade during chopping. Use your favourite-sized cook's knife or a Santoku knife if not.

Preparing marrow

For ease, use the largest cook's knife you have.

Stuffed roast marrow

Marrow can be prepared in much the same way as courgettes. If you purely want to stuff and roast the marrow, slice the marrow in half lengthways and remove the seeds as you would with a cucumber (see page 59), using a tablespoon instead of a teaspoon.

SLICING COURGETTES

Grip the courgette in one hand and tuck your fingers out of the way. Rest the tip of your cook's knife on the board and pulling back, cut into the courgette with a forward motion, using the full 'belly' of the knife. The angle shown here is the 'oblique' – you can, of course, also cut into straight rounds at 90° to the courgette. Discard the stalk.

COURGETTE WEDGES

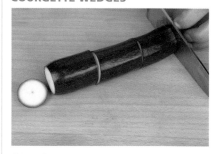

Step 1
Lie the courgette on the chopping board, then cut it into three or four cylindrical pieces 5 cm (2 in.) in length.

DICING COURGETTES

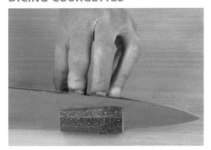

Step 1
Cut the courgette into cylindrical lengths as in step 1 of courgette wedges below. Take these cylinders and cut each into slices 1 cm (½ in.) in width.

Step 2
Take each slice and cut further into batons (the same width to match the cut in step 1). You can pile two or three pieces together and slice at once, although for more precision, cut one slice at a time.

Step 3
Take a pile of courgette batons, group them together, and using the claw method (see page 32), slice them into small dice.

Step 2
Stand the pieces up on their flat ends. Steady the zucchini with the thumb and first finger of one hand and, placing the knife just in front of these fingers, slice in half using a downward motion. Place each piece flat-side down on the board.

Step 3
Take each half and slice equally in half again, gripping the zucchini using the claw method.

ZUCCHINI FRITTI

For traditional 'zucchini fritti', cut courgettes into small batons (about half the size of that created in step 2 above). Alternatively, peel the courgettes into ribbons (sometimes called 'tagliatelle') and fry the same way.

Squashes

The squash family members are of varying sizes and unusual shapes. Generally harvested at maturity, they are then cured to further harden their skin.

Butternut squash and pumpkins are hard, dense vegetables, and as a result can be difficult to prepare. Always ensure your knife is sharp and that your chopping surface is steady. Unless you are stuffing or roasting two halves of squash, begin by cutting the vegetables into manageable sizes. This will make removing the skin, seeds and unwanted fleshy interior much easier.

Pumpkin is slightly softer than butternut squash, with less flesh and more pips. When preparing, it's important to match the size of the knife to the size of the pumpkin.

WHICH KNIFE?
Slicing, dicing and wedges

Use a very sharp, large cook's knife to tackle large squashes. For smaller squashes (such as an acorn squash, for example), a medium-sized cook's knife will do fine. Bear in mind the stalk is very tough to cut through.

Peeling a squash

Because the angles of a squash often vary, using a vegetable peeler can be testing. An alternative is to place smaller cut pieces of squash on their flat side and remove the skin using a cook's knife (see page 61).

DICING BUTTERNUT (SKIN ON)

Step 1
Place the squash onto its side. At the point where the long length of the squash meets the base 'bulb' shape, slice across. You should be left with a long piece and short bulbous piece.

Step 5
To prepare the bulbous part of the squash, slice in half through the seeds. Take a spoon and remove the seeds (discard, or roast and eat them). Then turn the squash piece onto its flat side and slice into even pieces. Group the pieces, turn 90°, then chop into dice as above.

Step 2

Take the longer piece, cut off the stalk and discard. Turn the squash onto its flat end and block-off (see page 47), to create a square piece of four equal sides.

Step 3

Cut this square piece into flat slabs of the desired thickness. Then take each piece and cut further into rectangles. (Try to cut into even pieces to get the most out of your piece of squash).

Step 4

Line these rectangles up and then cut into even final dice.

PUMPKIN WEDGES (SKIN ON)

Step 1

Place the pumpkin on the chopping board on its base. Holding the handle in one hand and using the heel of the hand near the tip of the knife in the other, slice in half. Cut downwards, all the way through the stalk and the main body of the pumpkin, using your body weight if necessary.

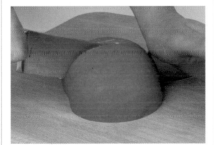

Step 2

Place the two halves, cut-side down, onto your chopping board. Slice each half in half again. Depending on the variety of pumpkin and its ripeness, it may be easier to wait until the end to remove the seeds, otherwise, do it now using a spoon (see step 5 opposite).

Step 3

Take each quarter and, holding on one side, cut into the desired thickness for a perfect wedge. Proceed carefully with this technique, as occasionally the wedge can snap under the pressure of the cut. Remove the seeds with a spoon or cut them out with a small kitchen knife, if you haven't done so already.

Cucumber

The refreshing cucumber can be prepared in a variety of ways. The juicy, moist texture is at its best just after it has been sliced. Because the cucumber seeds are particularly watery, many recipes call for them to be removed.

A cucumber should be handled delicately – a knife will easily glide through its soft texture. To prevent the moist flesh sticking to the knife, you may want to consider using a hollowed-edge blade if you have one. For a melt-in-the-mouth texture, peel the cucumber; for perfect, regular slices, cut slowly until you become more experienced. Alternatively, use a mandoline.

PEELING

Step 1
Trim the ends of the cucumber, then cut into manageable lengths (either in half or thirds).

SLICING INTO ROUNDS

Grip the cucumber in one hand using the claw method (see page 32), and with the classic rolling chop (see page 32), and a suitably sized cook's knife, slice into the required thickness.

WHICH KNIFE?
Slicing and dicing

For general cucumber preparation, use your favourite cook's knife or Santoku knife, although a hollowed-edge knife is effective in preventing the slices from sticking to the blade, so use one of these if you have one.

For smaller, more intricate work, use a small serrated knife.

Finely chopping a gherkin

For finely chopped gherkin, start by cutting the gherkin in half so that there are two short pieces. Standing each of these pieces on their flat end, slice into slabs of the desired thickness. Then, piling two or three of these flat pieces on top of one other, slice into long pieces. Turn the pile 90°, and slice again into dice (as shown in steps 1 and 2 of *Dicing Courgettes* on page 55).

Step 2
Take a piece into one hand and, holding a peeler in the other hand, peel all the way around. Leave space in between the peel for a ridged effect. Then either slice for a salad (see below), or follow on and deseed.

DESEEDING

Step 1
Halve a cucumber and cut each piece in half lengthways. Place each half on the chopping board, curved-side down.

Step 2
Take a teaspoon and, holding it at a 90° angle to the cucumber, dig the spoon into the centre. Gently drag the teaspoon from one end of the cucumber to the other (be careful not to dig too deep). The seeds will accumulate in the path of the spoon (discard them). You may need to go back over a second time to remove any excess seeds.

SLICING INTO HALF-MOONS

After deseeding (see steps 1 and 2 above), turn the cucumber flat-side down and slice across the width for curved, half-moon slices.

DICING

Step 1
After deeseeding, turn the cucumber onto its curved side. Slice into long thin lengths.

Step 2
Group the lengths of cucumber into manageable piles, using the claw method. Turn 90° and cut into dice. This method is ideal for preparing salsa.

Swede, celeriac and turnip

These root vegetables are bulky and firm, so handle your knife cautiously when implementing these knife skills.

As with most ingredients, try to find or make a flat surface as soon as possible to give you an extra steady support while cutting. The cutting steps here can be applied to any root vegetable.

PEELING

Step 1
These vegetables have tricky shapes and hairy, knobbly skin, so peeling with a peeler isn't a good option – peeling with a knife is much more effective. Start by topping and tailing.

WHICH KNIFE?
Peeling and dicing

A medium or large cook's knife is essential for the preparation of these sturdy root vegetables. A Santoku blade is also a good consideration.

For wafer-thin slices

If you have practised knife skills, use a large cook's knife to slice into wafer-thin slices. It is easier to cut a smaller piece, so try cutting a whole swede in half, to make it more manageable. For beautiful, even, thin slices of celeriac, turnip or swede every time, use a mandoline. You may need to cut the root vegetable to fit the width of the mandoline.

For thin julienne

Using a cook's knife, take a pile of thin slices and cut into thin julienne.

To prepare a baby turnip

While peeling the turnip, try to keep the green root intact for attractive presentation. The best way to achieve this is by using a peeler to peel the skin in sweeping, even movements. You can also use a turning knife here, as the curve of the blade is ideal. Keep the pointy base on, unless you require the turnip to stand up on the plate.

DICING

Step 1
Lay the peeled vegetable onto its flat base and insert your knife. Use the claw method (see page 32) to grip.

Step 2
Stand on a flat end and place the knife just inside the skin. Cut down, being careful not to remove too much flesh.

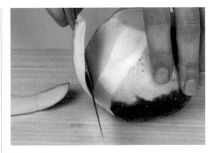

Step 3
Work with the curve of the vegetable, from the top to the bottom. Take off just enough skin to expose the flesh.

Step 4
Turn the vegetable slightly and continue this process until all of the skin has been removed. The same rules can be applied to turnips (which are generally smaller and smoother than other members of the tuber family), and celeriac. Note that celeriac has a particularly knobbly texture and so you may have to remove more skin.

Step 2
Begin cutting into slices and continue along the width of the vegetable, cutting at intervals suitable for your recipe. Attempt to keep slices even.

Step 3
Lay each piece on its largest flat side, and cut into thick strips as shown.

Step 4
Group the pieces together, turn 90° and chop into the required sized dice.

Leafy vegetables

Removing the core or stems are the main tasks required for the preparation of leafy green vegetables.

A favoured technique for preparing lettuce is to simply use your hands and tear it. This enables the lettuce to rip along its natural cell lines without damaging the leaf and causing early deterioration. Using a knife causes bruising, however, cutting lettuce is sometimes appropriate – in restaurants Caesar salads are often chopped with a knife just prior to serving.

Spinach requires little preparation (if any at all) other than removing the stalk if the spinach is older than baby spinach. This can be done by hand too, but hardier leafy varieties (such as kale) may need to have their woody stems removed with a knife. Finely chopped stems are occasionally used for the contrasting texture that they add. With Swiss chard and pak choi, the stem is (despite its bitterness) often used.

ROMAINE LETTUCE

Step 1
Trim the lettuce at the bottom. Collate a manageable amount of leaves and roll into a cigar shape.

PAK CHOI

A very quick vegetable to prepare, simply trim the base of the pak choi to release the leaves. It is generally preferred to leave the leaves whole. Serve the stalk crunchy, as the leaves cook a lot quicker than the stalk.

WHICH KNIFE?
The stalks and leaves

A medium or large cook's knife is perfect for removing the stalks and chopping the leaves of these vegetables. If you have a Santoku knife, use that.

Cutting lettuce

Ideally, use your hands to tear lettuce. If you do need to use a knife, use a plastic lettuce knife or a ceramic knife, as they will not cause the lettuce to turn brown.

Step 2
Using the rolling chop (see page 32), and the claw method to grip, slice the lettuce into the desired width to give an attractive shredded effect.

KALE

Step 1
To remove the woody stem, first trim the kale at the base to release the stalks.

Step 2
Take one stalk at a time and cut along the side of the stalk to separate it from the leaves, and remove. You can do this easily with your hands, although it is much easier with a knife. Either leave the leaves whole, or slice as desired.

SWISS CHARD

Step 1
The stalk of Swiss chard usually needs longer to cook than the leaves, so begin by separating the two.
  Using a medium cook's knife, place the chard leaf on your chopping board. Steadying the stem in one hand, cut the stalk away in a 'V' shape. Repeat this for all of the leaves.

Step 2
Chop the stalk, using a rolling chop and slicing as finely as required.

Step 3
Take the leaves and, rolling them up into a cigar shape, chop using the rolling chop technique as you would for Romaine lettuce.

Potatoes

Purple, red-skinned, sweet or plain, this versatile tuber is one of the most popular vegetables in the world, and can be prepared and used in a variety of ways.

Floury potato varieties such as Golden Wonder are best for mashed potatoes, baked potatoes, chips and roasting. Waxy varieties such as the Charlotte potato are best for eating in salads. Maris Piper varieties work well with any dish. Whether you choose to peel them or not depends on personal preference and the preparation methods called for in the recipe. To prevent the potatoes from turning brown after peeling, place them in cold water until you are ready to use.

PEELING

Using a vegetable peeler, peel potatoes from one end to the other, turning when necessary. Always peel over a bowl of water, dipping the potato into the water regularly as you peel – this prevents dirt becoming ingrained in the potato.

WHICH KNIFE?
Peeling, slicing or dicing

As the potato is a starchy vegetable, a hollowed-edge knife is perfect for preventing the potato from sticking to the knife. Otherwise, a medium cook's knife or Santoku knife works fine.

Potato crisps

To make your own potato crisps, follow the dauphinoise procedure shown opposite, but ensure slices are very thin and as even as possible, so they cook at the same time. If you have a mandoline, use it.

Sweet potato

Sweet potatoes can be treated in the same way as regular potatoes and are particularly great on the barbeque in slices, or baked whole in the oven. The skin is thin, so you may want to keep it on during cooking.

POTATO WEDGES

Start by cutting the potato in half. Place each half of the potato onto its flat side. Using your cook's knife at an angle, cut the potato into thirds or quarters, through the skin. Each half should produce three or four wedges.

SLICING FOR DAUPHINOISE

Restaurant chefs often use a mandoline to produce thinly sliced potatoes. If you don't have access to a mandoline, you can use a grater with a suitably wide – and sharp – cutter on one side. Otherwise, use a sharp knife.

Steady the potato as much as possible with one hand using the claw method (see page 32). Cut thinly and slowly, into slices, making sure that the cut is straight so the potato will cook evenly.

PREPARING FOR ROASTING

Step 1
Peel the potatoes over water. Using the under-the-bridge method (see page 30) to steady the potato, cut in half.

Step 2
Turn each piece onto its flat side and cut in half again. If the potatoes are large, repeat the procedure and cut in half again. Either way, ensure that the pieces are uniform in size, but not too small (especially if you are blanching them first).

CHUNKY CHIPS

Step 1
Start with a whole peeled potato. Block-off (see page 47) the potato by cutting off the curved edges to give a rectangular shape.

Step 2
Slice the rectangular potato into wide-slabbed pieces (the desired width of your chips).

Step 3
Take each of these pieces and cut into individual chips. Store in water until cooking. For a more rustic finish, leave the skin and curved edges on the potato instead of blocking-off, and just cut into rectangles as much as possible.

Tomatoes

There is nothing tastier than a ripe, juicy, tart and sweet tomato. A good tomato should feel heavy for its size and have a smooth and shiny skin.

To protect their wonderful taste, store tomatoes at 9°C (48°F) – storing in the fridge kills their flavour. Removing the tomato skin opens up lots of possibilities and offers lovely, uninterrupted tomato flesh. For some classic tomato sauces, all that is used is the flesh (no skin and no seeds), so peeling is an important skill to learn. Consider using a specialist serrated tomato knife – tomato skin will easily blunt a regular knife.

WHICH KNIFE?

Peeling

To remove the skin after blanching, use a turning knife, or a kitchen knife.

Slicing, deseeding and concasse

A serrated tomato knife is ideal for all tomato preparation. Once the skin is removed any Santoku knife or medium cook's knife can be used for cutting the flesh, although a serrated tomato knife or fruit knife is preferable.

For diamonds

Cut tomato strips on the cross (see step 2 of Slicing, page 47) for an elongated, diamond shape variation.

PEELING A TOMATO

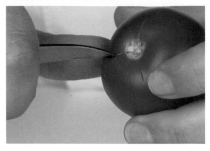

Step 1
Make a small cross incision at the top and base of the tomato, using a turning knife. Be careful not to cut too deep into the flesh, just cut through the skin to allow it to peel back slightly during the blanching process.

DESEEDING AND CONCASSE

Step 1
Concasse preparation is an arduous task, but the resulting delicate, small squares of tomato are perfect for dressings, salsas and garnishes.
   Place a tomato on a chopping board. Using a serrated knife and slicing downwards, quarter the tomato.

Step 2

Dip the tomato in boiling water for 10–15 seconds, depending on ripeness. The skin should start to peel back (and if it doesn't, simply cook for a little while longer). As the skin begins to peel back, remove immediately and plunge into cold water to prevent further cooking.

Step 3

As soon as the tomato is cold (after being submerged for 5 minutes), remove from the water and drain on a kitchen towel. If you are submerging a batch of tomatoes, change the water as soon as it starts to heat up. Secure the back of your knife under the tomato skin and peel back to remove.

SLICING

Gripping the tomato with one hand and using the claw method (see page 32), slice using a serrated tomato knife or serrated fruit knife.

Step 2

Deseed the tomato quarters using a teaspoon or a serrated knife (pictured here). Begin by placing the tomato on its curved side. Insert the knife and cut at an angle, away from you, to remove all of the seeds.

Step 3

Place each tomato 'petal' curved-side up on your chopping board. For perfect square concasse, it is best to block-off (see page 47) first by trimming off all the curved edges. Cut into 1 cm (⅖ in.) strips.

Step 4

Grouping a few slices together at a time, slice the strips into perfect concasse. The pieces should be about 1 cm x 1 cm (⅖ in. x ⅖ in.). For a slight variation, cut the strips into diamonds (see opposite).

Peppers and chillies

Peppers and chillies are versatile vegetables that add great colour to many dishes, and are equally tasty raw or cooked.

With both vegetables it is easier to cut through the flesh side first, as the skin is as tough as that of a tomato or aubergine, although, of course, this is not always possible in the preliminary preparations. Peppers are ideal for stuffing, especially in vegetarian or vegan dishes. They also provide the perfect portion size. Served either in open halves (Mediterranean style) or standing up (North-African style), stuffed peppers are simple to prepare.

WHICH KNIFE?
General preparation

For general preparation, a medium cook's knife or Santoku blade works well, once you have cut through the skin. For cutting through the skin, use a serrated knife (tomato knife or fruit knife) – the serrations tackle the tough skin well.

Roast whole or in pieces?

Cutting a pepper into pieces prior to roasting is arguably the tidier method, as the seeds are removed at this point. Also, pieces don't need to be turned regularly, unlike a whole pepper, which does.

PREPARING LOBES FOR ROASTING

Step 1
A popular method of cooking peppers is to roast them. You can either leave them whole to do this, or cut them into their natural 'lobes'. Begin by slicing along the natural lines with the 'belly' of your knife. There will be three or four natural lines depending on the variety.

SLICING

Step 1
Following on from step 4 above, take each lobe of pepper and block-off (see page 47) so you have a regular oblong shape. Then cut this blocked-off piece in half so you have two equal pieces approx 2.5 cm x 2.5 cm (1 in. x 1 in.).

Step 2
Using your fingers, pull the lobes away from the main body of the pepper.

Step 3
Remove the stalk and seeds from each of the lobes and discard. Try to remove the seeds in one piece – they have a tendency to scatter.

Step 4
Trim away the bitter white membrane from each piece – try to retain the natural shape as much as possible. Cut each piece in half if necessary, and you are now ready to roast.

Step 2
Next, slice the lobe into strips approximately 1 cm (⅖ in.) in width.

DICING (FOR KEBABS)

Following on from step 1 of Slicing, take a blocked-off lobe of pepper and cut in half, so you have two equal pieces measuring approximately 2.5 cm x 2.5 cm (1 in. x 1 in.).

DICING (FOR SALSA OR SALADS)

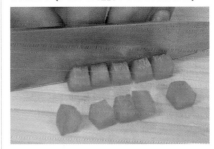

Following on from step 2 of Slicing, group a pile of pepper slices together, turn 90° and slice again into perfect 1 cm (⅖ in.) squares.

PREPARING FOR STUFFING: MEDITERRANEAN STYLE

Step 1
Place the pepper on its side. Before slicing, estimate where best to cut the pepper in order to produce two evenly sized pieces. Grip the pepper between your fingers and slice down using a large cook's knife. Try to slice equally through the green stalk.

Step 2
Cut away the seed core and remove any fallen seeds from each half of the pepper. Keep the stalk intact. Also, trim away any bitter white membrane from the flesh.

Step 3
Stuff each half according to your recipe.

CHILLIES

Chillies are the most commonly grown spice in the world. It is difficult to anticipate how hot an individual chilli will be, as even those derived from the same plant differ. Unless it's being used in a marinade or to flavour stock, a chilli needs to be finely chopped so that the pungency is evenly distributed.
Note that chillies need to be handled with care. Use gloves, wash your hands vigorously with hot soapy water, or follow the fork method opposite.

CHILLI RINGS

Using a small, serrated knife or a medium cook's knife, slice down through the chilli to create a ring. Continue slicing as thinly as you like, all the way up the chilli. Note, you can also do this on the cross (see step 2 of Slicing, page 47) for a slanted effect.

FINELY CHOPPING CHILI

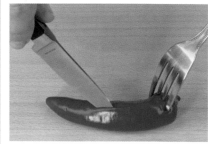

Step 1
Place the chilli on your chopping board. Pierce with a fork at one end to hold the chilli steady, and use a knife to cut in half lengthways. Trim off the root.

PREPARING FOR STUFFING: NORTH-AFRICAN STYLE

Step 1
Place the pepper on its side and position your knife roughly three-quarters of the way up the pepper. Cut all the way through, removing the top of the pepper.

Step 2
Cut through the membrane that attaches the seed core to the pepper and remove with your fingers. Remove any fallen seeds and trim away any excess white membrane.

Step 3
If the pepper won't stand upright, shave a little off the bottom of the pepper to create a level base. It is now ready for stuffing.

Step 2
Holding the root end of the chilli with the fork, take a teaspoon and gently drag the seeds and any white membrane out of the chilli. Note that the seeds and membrane harbour most of the pungency.

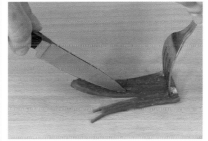

Step 3
Retain the fork's hold of the root end of the chilli. Using the tip of a small kitchen knife (or a cook's knife), make an incision near the top of the chilli and drag the knife down to the tip, trying to keep in a straight line. Repeat all the way across the chilli.

Step 4
With a cook's knife, use a rolling chop (see page 32) to cut the chilli into very fine dice.

Aubergine

With its spongy texture and pleasantly distinct taste, the versatility of aubergine knows no bounds. It adapts well to many different styles and flavours.

Whatever shape, size or colour you choose, the perfect aubergine should be shiny, smooth and blemish free. Aubergines respond well to a serrated knife – the serrations of the blade effortlessly tackle the tough skin, yet are gentle to the delicate flesh of the interior. Use a stainless steel rather than a carbon steel knife – the carbon reacts with the aubergine's nutrients and causes it to turn black.

WHICH KNIFE?
General preparation

For aubergine preparation, use a serrated knife. A long one is ideal (so if you only have a bread knife use it here). Once through the skin, however, a regular cook's knife or Santoku knife will work well. For super thin slices, use a meat slicer or mandoline if accessible.

Salting the aubergine

You may consider salting your aubergine to remove excess moisture and bitterness. This is also known as 'sweating' the aubergine. Salting is also useful to prevent the aubergine from absorbing too much oil during cooking. Remember to pat dry before cooking.

SLICING LENGTHWAYS

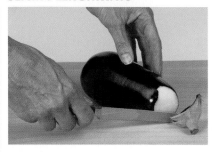

Step 1
Place the aubergine on its side and slice the stalk off. Cut as near to the stalk as possible. Discard the stalk.

CUTTING INTO ROUNDS

Place the aubergine on a chopping board and remove the stalk as shown in step 1 above. Gripping with one hand using the claw method (see page 32), slice through the aubergine to create rounds of the desired width. Continue along the length and discard the stalk.

Step 2

Using a serrated knife, carve through the aubergine lengthways to produce even slices. Steady the aubergine as well as you can with one hand while cutting with the other.

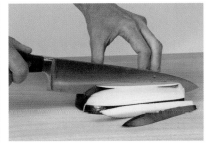

Step 3

Continue for long strips: select a small pile of aubergine slices (perhaps 3). First place the pile flat-side down, and then slice or carve downwards into 1 cm (⅖ in.) slices.

DICING

Follow steps 1–3 for slicing lengthways. Take a group of aubergine strips, turn 90°, and using a rolling chop, cut the strips into small dice.

PREPARING FOR STUFFING

Step 1

Lay an aubergine on its side. Take a large serrated knife if you have one or a plain cook's knife, and using the under-the-bridge method (see page 30), insert the knife as centrally as you can and cut downwards. Aim to cut precisely through the stem and through to the board.

Step 2

Lay the two halves curved-side down on the board. Taking one half at a time, use a spoon to pierce the soft flesh and scoop to remove it from the aubergine. Aim to leave about 1 cm (⅖ in.) of flesh attached to the skin.

Step 3

Take the flesh and chop using a cross chop (see page 34) and add to your favourite stuffing recipe. Note that the flesh will turn brown quickly.

Turning vegetables

Turning is a classic French method of preparing vegetables. Root vegetables are especially well suited to turning because they are even and solid.

A turned vegetable should be a perfect barrel shape, making it very pleasing to the eye. To do this properly, use a turning knife. To achieve the authentic barrel shape it is necessary to hold both the vegetable and the knife in your hands (no leaning on a chopping board here, so be careful). It isn't essential to 'block off' the vegetable first, but in doing so, it is easier to maintain consistency.

WHICH KNIFE?
General preparation

The turning knife is ergonomically shaped to enable perfect curved results, so make this knife your first choice. Turning is possible with a kitchen knife, but once you are used to a turning knife it is difficult to revert to a regular shaped blade. Turning knives are pretty useful to have around and inexpensive, so they are a good investment.

Turning tips

Visualize the desired end result before you begin. Try to work as evenly as possible – this process requires a lot of practice to achieve consistency.

You can also turn vegetables after they have been cooked – since they are softer, they are then easier to cut. If a recipe requires the turning to be done raw, make sure the vegetables are at room temperature, to aid cutting.

TURNING CARROTS

Step 1
In order to start with an even shape, block off the carrot to create a rectangle shape. Hold the vegetable in one hand between the thumb and first finger. With the turning knife in the other hand, start cutting from the top of the vegetable, close to the centre. Cut downwards, towards yourself, slowly.

TURNING COURGETTES

To turn courgettes, begin by cutting into lengths, measuring four fingers. Then cut into quarters. Turn each quarter following steps 1–4 above. Courgettes are challenging to turn if you are trying to maintain the green edge, and often the turned vegetable ends up flat rather than round. Just aim for consistency.

Step 2
Curve out and then curve in, as uniformly as possible.

Step 3
Continue to turn all sides of the carrot.

Step 4
The result should be consistent, smooth and curved barrels.

TURNING POTATOES

Step 1
Start off with a blocked-off piece of potato. Using the turning knife, start at the top of one corner and glide downwards and outwards at the same time (see step 1 above). For support, keep the thumb at the potato base.

Step 2
Once you reach the central part of the potato, start to curve the knife back in, at the same angle used to begin. Try to keep your wrist flexible to accommodate turning the vegetable up and down, and side to side.

Step 3
Continue around the vegetable until perfect, evenly turned faces are achieved.

Mushrooms

The soft consistency of mushrooms makes them perfect for practising the hammer-style chop (see page 30).

Try to avoid washing mushrooms, as they will only absorb the water, causing them to steam rather than brown during cooking. The best way to clean a mushroom is to wipe or brush away any dirt. Specialist mushroom brushes exist for this purpose although a new pastry brush is fine for this task.

Mushrooms can stick to the knife during the cutting process, but using a hollowed-edge knife is useful in preventing this.

SLICING

Step 1
To remove the core, hold the side of the mushroom and slice away from you, cutting as close to the base of the stalk as possible. Alternatively, gently pull off the core with your fingers.

CLEANING WILD MUSHROOMS

The many small 'nooks and crannies' in some wild mushrooms (chanterelle is pictured here) mean they tend to harbour lots of grit. Clean all of this away, without letting them come into contact with water. Use a brush to clean as much of this away as possible.

WHICH KNIFE?
General preparation

A regular cook's knife or smaller is ideal for mushroom slicing, and a hollowed-edge blade will prevent mushrooms from sticking.

For the mushroom garnish

Use a turning knife to craft the white cap mushroom garnish (see opposite).

Slicing truffles

To get the most out of rare truffles, slice them thinly using a special truffle cutter (see page 20) or a mandoline. Simply 'rub' the truffle back and forth – watching your fingers.

For a finer chop, take a pile of thin slices and cut further into strips. Turn 90° and cut again into tiny brunoise (see Glossary, pages 172–173).

FINE DICE OF MUSHROOM (DUXELLE)

Step 2
Turn the mushroom flat-side down and, using the claw method (see page 32), slice downwards through the mushroom, as thinly as required.

Step 1
Follow steps 1 and 2 for slicing. Create a pile of mushroom slices. For very fine dice, cut in half again, lengthways. Make further cuts lengthways to slice the mushroom pile into thin strips, measuring 5 mm (⅕ in.).

Step 2
Group the mushroom slices, turn 90° and slice again into fine dice.

PREPARING WILD MUSHROOMS

MUSHROOM GARNISH

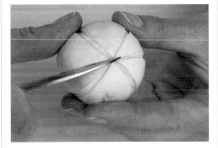

Once cleaned, use your fingers to pull apart the mushroom into two pieces. Using your fingers will give a more organic shape, although this is not always possible with some wild mushrooms (such as porcini). If this is the case, use a kitchen knife to divide as desired.

Step 1
For a refined French garnish for mushrooms, use a turning knife to carve the mushroom cap. First make an incision in the centre of the mushroom top, and curve the knife around the edge to the bottom. Cut at a slight angle and dig out a small 'V' during the process.

Step 2
Continue making these cuts all the way around the mushroom, cutting across the intersecting lines for an attractive finish, great for garnishing.

Ginger

Ginger is a useful, knobbly spice with a hot and powerful taste. Most recipes will require a measurement of ginger (2.5 cm [1 in.], for example) or a precise weight.

You can peel ginger with a peeler but a more efficient way is to block-off with a knife. This will leave a neat, uniform shape, ideal for preparing julienned ginger from. A more economical – and unexpectedly productive – method of peeling ginger is to shave off the skin using a teaspoon. This technique enables maximum use of the flesh with minimal waste.

PEELING WITH A KNIFE

Take the ginger in one hand and the turning knife in the other, and peel just under the skin. The turning knife should glide over the lumps and bumps relatively easily.

JULIENNE OF GINGER

Step 1
Take a blocked-off piece of ginger. Grip the ginger with one hand using the claw grip and slice lengthways into very thin slabs.

WHICH KNIFE?
Peeling and slicing

A small kitchen knife or turning knife is perfect for peeling around the knobbly edges. For julienne and other slicing, upgrade to a medium cook's knife or Santoku knife. A Chinese cleaver is also a good design to get very fine julienned slices.

Grating ginger

An alternative to fine dicing is to grate ginger. Ginger can also be grated from frozen, although this does change the end result slightly, leaving it more watery than if it were fresh.

Using trimmings

Why not slice any trimmings thinly and make a refreshing ginger tea? Simply pour boiling water over the trimmings and brew like regular tea. You can of course use ginger that is beautifully sliced, too.

PEELING WITH A SPOON

Take the teaspoon in one hand and grip the ginger in the other (between thumb and first finger). Using the same hand that is holding the spoon, rest your thumb against the ginger and scrape the ginger skin downwards – it should come away easily. Rotate the ginger and repeat until peeled.

Step 2
Take a pile of about three ginger slabs and, still using the claw grip to steady, slice very thinly into julienne.

BLOCKING-OFF

Place the ginger on your chopping board and cut off the curved edges using a medium cook's knife. Block-off the ginger into a rectangle. Aim for a piece that is as even as possible, without having to cut too much off. Discard the skin or see panel opposite on using trimmings.

DICING

For fine dice, take a pile of julienned ginger. Grip using the claw method, turn 90° and chop finely using the rolling chop (see page 32).

SLICING INTO ROUNDS

Place the peeled, blocked-off ginger on a chopping board. Find its flattest side. Grip in one hand using the claw grip (see page 32) and slice into rounds of the desired width.

Galangal

Similar in flavour and appearance to ginger, fragrant galangal can be difficult to cut as it is firmer and woodier than ginger. Trim off the firm shoots with a medium cook's knife or Santoku knife. Just use the young and tender insides, either finely chopped or sliced.

Beans: runner and French

Most beans need little preparation other than podding (broadbeans, for example) or topping and tailing prior to cooking.

Topping simply means removing the connecting stem with which the bean attaches itself to the main plant. Tailing means to trim the 'tail', usually a curly end to the bean, although this is only necessary in certain styles of cuisine. Runner beans mostly need the sides removing as they can be stringy and hard to digest. To test, carefully put your knife through the top of the bean without cutting right through, then pull downwards. If a thick thread comes away, the beans need destringing, so do the same on the other side.

Topping and tailing

To trim beans, use a medium-sized cook's knife or Santoku blade.

Julienne

For more precise chopping, a shorter blade like a small cook's knife, Santoku knife or even a Chinese cleaver gives more precision.

Removing the tail

Do we need to remove the tail? It is harmless and certainly edible. Some styles of cuisine would encourage refined preparation by removing the tail (i.e. French), whereas others would prefer leaving the tail on (rustic Italian), for a much more natural approach.

TOPPING AND TAILING

Step 1
Group the beans together in a pile, and hold them together using the claw method (see page 32). Use the rolling chop (see page 32) to chop off the tail as close as you can to the end.

LONG JULIENNE OF RUNNER BEANS

Step 1
These long, elegant cuts require precision chopping. Firstly take a runner bean and top and tail it. Then cut the beans into 5 cm (2 in.) long batons.

Step 2

To 'top' the beans, turn them around the other way. Make sure they are all grouped together at one end and trim the beans in the same way as in step 1.

Step 3

Now you can either leave the beans whole, or alternatively cut in half into smaller 'batons', as shown here.

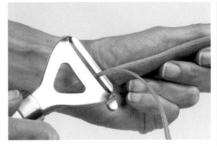

A vegetable peeler is an ideal tool to use as it only removes a minimum amount of bean.

Take the bean in one hand and, with a peeler in the other hand, gently peel the bean on one side. Turn the bean over and peel the other side.

Step 2

Grip each baton as best you can with one hand, using the claw method – this is difficult as the beans are so thin. Then, slice thinly and as uniformly as possible into julienne. This cut is ideal for using beans as a garnish.

DIAMOND CUT

Take a trimmed and destringed runner bean and place flat on your board. Turn your knife to a 45° angle to the bean and, gripping the bean in one hand using the claw method, slice on the cross (see step 2 of Slicing, page 47) at 2.5 cm (1 in.) intervals. Maintain the 45° angle all along the bean.

THIN DIAMOND CUT

For a finer finish, slice the beans into a very thin julienne using the same angle as the diamond cut. As this slice is very fine, destringing isn't necessary.

Artichoke

The artichoke is unlike any other vegetable in its tricky preparation techniques – preparing the elegant artichoke is very much a labour of love, but worth the effort in the end. The larger the artichoke grows, the tougher the consistency becomes, so ensure that your knife is very sharp.

You can prepare the artichoke in a number of ways – the easiest is to cook and serve the artichoke whole (with no preparation). The artichoke is then 'prepared' by whoever is eating it: first by pulling off the leaves (and sucking out the flesh), then by scooping out the hairy choke and finally eating the heart. Preparing the artichoke before serving leaves just the heart, as shown in steps 4 and 5 opposite. When preparing an artichoke for the first time it will feel like you are throwing most of it away – unfortunately, there is a lot of waste.

Step 1
Place the artichoke on its side. Grip the large end in one hand and use a large serrated knife or large cook's knife to trim off the woody stem.

Step 5
Next, we need to remove what is called the 'hairy choke' (this is not necessary in younger artichokes).

Cut off the excess leaves, then take a spoon and scoop the hairy choke out. There will be quite an abundance of this matter. It is not edible, so discard. The artichoke heart is now ready to cook. Either place immediately into acidulated water or the chosen cooking liquid, or slice or quarter and then cook.

WHICH KNIFE?
General preparation

Here we have used a serrated bread knife that tackles the rigid exterior well. A turning knife is also useful for trimming any edges (step 4). The skin is slightly softer on smaller artichokes, so a medium cook's knife will do.

Preservation

Remember that when artichokes are cut, their exposed sides will turn brown (oxidize) quickly, so either rub the exposed sides with a lemon half, or place the artichoke into some acidulated (lemon) water.

Step 2
Shave off a small amount of the bottom of the artichoke globe to leave a flat, even base.

Step 3
Turn and carve around the edge of the artichoke, removing the tough outer leaves as you work. Try not to remove too much flesh.

Step 4
Using a turning knife or small kitchen knife, trim a small amount off the base of the artichoke.

PREPARING BABY ARTICHOKES

Step 1
Young artichokes are much easier to prepare, as the flesh is still tender. The hairy choke is less developed and edible so it doesn't have to be removed. Place the artichoke on its side and remove the tough outer leaves. Turn the artichoke and repeat to remaining sides.

Step 2
Slice off about one-third of the leaves from the end of the artichoke, and discard them.

Step 3
Next, peel the stalk. Take the artichoke in one hand and a turning knife in the other and gently carve off the tough outer skin. Take off as little as possible. Be sure to follow the natural lines of the artichoke stem. Thinly slice the remaining flesh and eat raw or deep-fried, or, for a simple cooking method, cook whole.

Herbs

With a wealth of different flavours and aromas to choose from, there is nothing better to brighten, lift or enhance a dish than freshly chopped herbs.

One of the most enjoyable things about chopping herbs is the scent released during chopping. Large bunches bought at farmers' markets are cheaper and invariably have an abundance of flavour compared to plastic sealed tubs bought in supermarkets. Growing pots are an option but are only designed to last a few weeks. There are a number of knife skills required for the preparation of herbs, all of which are discussed in the coming pages.

WHICH KNIFE?
General preparation

The mezzaluna is fantastic for cross-chopping and offers a thorough, even chop of parsley and coriander. For all other herbs, use your sharpest cook's knife (medium or large). If the knife isn't sharp enough, the herbs will darken and squash on the edges. For precision chopping, make sure the chopping board is completely flat.

Coriander

The refreshing aromatics of coriander are irreplaceable. Most of this unique flavour is concentrated in the roots and stalks. The stalks are mostly used in curries or soups and removed just before eating. As the stalks are quite soft, it is acceptable to chop them finely and use with the leaves. The leaves can also be picked separately and used in the same way as parsley leaves in Asian salads, or as a garnish.

PARSLEY: PREPARING FOR SALADS

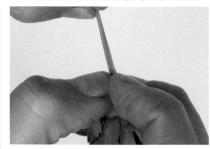

Only the tender leaves are required for salads, so the stalk needs to be removed. There is a lot of flavour in parsley stalks however, so you may want to use them to flavour stocks or soups.

Hold the stalk in one hand and, with the other, firmly pull the leaves away.

CORIANDER: ROOTS AND LEAVES

Step 1
Grip the coriander in one hand using the claw method (see page 32). Starting at the root end, begin slicing the roots.

PARSLEY: TEARING

For a relaxed, rustic feel, tear the parsley leaves with your hands instead of chopping them.

Pick the leaves in the same way as for a herb salad, and tear between your fingers into the desired size.

PARSLEY: CROSS-CHOPPING

Step 1

Parsley is a hardy herb that responds well to cross-chopping. Remove the stalks and place the parsley on the chopping board. Holding the mezzaluna with both hands, roughly chop the leaves with a backward and forward motion to ensure even chopping.

Step 2

If fine parsley is required, continue chopping – tossing the herbs back on themselves as you go – until a finer chopped result is produced. The mezzaluna is ideal, although a cook's knife can be used to the same effect.

Step 2

Continue cutting along in the same motion. When you reach the leaves (where the stalk inevitably gets finer) simply carry on slicing in the same way, as fine as you wish.

CORIANDER: TENDER LEAVES ONLY

Step 1

Pick the coriander leaves from the stalk. Start to cross-chop (see page 34) by holding the tip of the knife in one hand and the handle in the other. Move the handle around up and down, and back and forth, over the herbs, keeping the tip in the same place, so you are using it as a sort of pivot.

Step 2

Scrape the herbs up using the belly of the knife and redistribute them in a pile. Continue to cross-chop until the herbs are as finely chopped as desired.

Mint

Prepare mint in the same way as for basil (see opposite). You can also tear mint for a more rustic feel. For a mint garnish, simply pick the small beautiful buds and store in water until needed.

Sage

Pick perfect small sage leaves and use them whole. For larger leaves, roll into a cigar (as shown in basil julienne opposite).

Chervil

With the appearance similar to that of flat parsley only smaller, the beautiful leaves are the perfect delicate garnish. For chopped chervil, pick the leaves and handle in the same way as dill (see opposite).

Tarragon

The tender leaves of tarragon should be picked from the woody stalk and chopped in the same way as dill (see opposite).

Oregano and marjoram

These herbs are difficult to tell apart. When picked, the leaves turn black quickly, so pick just prior to using. To chop them, follow the dill steps shown opposite.

DILL

Step 1
Take one stem of dill at a time. Hold by the stem and pull the tender leaves away. Try not to take any of the stem away (although it is easy for it to snap, so be careful). Repeat until you have used all the dill, then discard the stalks.

Step 2
Bunch the tender leaves together and hold with the claw grip (see page 32). Start with a rolling chop (see page 32) and roughly chop the leaves. Then, cross-chop (see page 34) and keep turning the leaves back on themselves, until all the leaves are chopped.

CHOPPING CHIVES FINELY

Step 1
An important knife skill, chopping a bunch of chives illustrates how good a chef's knife skills really are. If the chives are held together with an elastic band, leave this on – keeping the stalks together makes chopping easier. If the knife you are using isn't very sharp, the leaves will bruise (and turn dark green).

Step 2
Use a very sharp cook's knife or Santoku knife for this technique. Place the chives on a chopping board. Holding the chives with one hand, chop the chives using a rolling chop. Go as slowly as you need to get the perfect fine slice. If you require longer chive lengths, simply use a pair of kitchen scissors to snip them.

BASIL JULIENNE

Step 1
Basil can simply be torn, but for a more precise, even flavour, julienne or shred the basil. This method can also be applied to mint and sage.

Pick the basil leaves and discard the stalks. Take three or four leaves and pile them on top of each other.

Step 2
Roll the pile of leaves into a cigar shape and secure with one hand on your chopping board.

Step 3
Using the claw method to secure the leaves, and a rolling chop, carefully shred as finely as required.

ROSEMARY

Step 1
Hold the stalk in one hand and pull the leaves back on themselves to remove. They should come off easily, but do this gently so the stalk doesn't snap.

Step 2
Continue until all the leaves have been taken off the main stalk. Pile the rosemary leaves up.

Step 3
Chop the rosemary using the rolling chop. Try to grip the rosemary while you are doing this, using the claw method. You can cross-chop, although the rosemary will turn black quickly.

fruit

Most fruit can be eaten in its raw state with little or no preparation at all, although there are a number of preparation methods required for a more professional finish – for fruit garnishes or pastry recipes, for example. This chapter highlights the most useful knife skills, from the basic portioning of a grapefruit, to the more complex mango porcupine.

Fruit knives

A large range of knives is available to suit the array of unusual shapes and sizes of fruit. A serrated knife is most commonly used for fruit as it tackles tough skins without damaging delicate interiors. Match the size of the knife to the fruit – a pineapple, for example, should be prepared with a large serrated knife.

Small ceramic knife

The perfect knife for intricate, delicate fruit work, this small ceramic knife is incredibly sharp and won't leave a residue on the ingredient being cut, like steel knives can. An apple prepared using this knife will not oxidize as quickly as that prepared with a steel knife.

Small serrated fruit knife

A fantastic all-purpose knife, great for cutting all kinds of fruit. Especially effective when slicing lemons and limes, this knife will always give great results. Like all serrated knives, in time the indented edge will become lost and so will eventually need replacing.

Cook's knife/Santoku knife

These all-purpose knives are ideal for preparing all kinds of fruits. Small, medium or large, choose your knife according to the fruit you are preparing and that which you find the most comfortable to work with. Pictured here is a Santoku knife with a hollowed edge.

Bread knife

The sharpness of the bread knife's indentations works exceptionally well on large fruits with tough exteriors (such as melons) and is ideal for making long, delicate and precise slices.

Kitchen knife

This small-scale knife is an indispensable kitchen tool and enables complete control to be exercised. Use it for small dice, or for peeling a variety of fruit and vegetables by hand.

Grapefruit knife

Unique in its appearance, the blade of this knife is specially designed to remove the flesh of a grapefruit from the pith. A great tool for portioning, this is obviously very useful if you eat a lot of grapefruit.

Turning knife

The curved blade of the turning knife is extremely effective for peeling fruit such as apples and turning them (see pages 74–75) should your recipe require it. The small blade offers greater control, making it perfect for coring fruit too.

Other useful tools

MELON BALLER/PARISIAN SCOOP
For sculpting spheres and deseeding pears (see page 19).

ZESTER
For obtaining the perfect pith-free citrus zest (see page 18).

CANELLE KNIFE
Also for pith-free zest, peel and fruit ribbons (see page 18).

PEELER
For a quick, easy and economical way to remove the skin from your fruits (see page 20).

Avocado

The creamy flesh of an avocado is hidden beneath its thick skin. Try not to handle an avocado too much during preparation as the edges will start to break down and the avocado will appear tired and unappetizing.

Depending on the variety, avocados can be peeled with a vegetable peeler (as with thinner green skin such as Fuerte) or the skin can be removed by hand (as with darker, bumpier skin, such as Hass). If using a peeler, simply take the avocado in one hand and peel with the other.

WHICH KNIFE?

General preparation

Any good cook's knife (medium or large) or Santoku knife works well. As the flesh is slightly sticky, a hollowed-edge knife is great in preventing the avocado from sticking to the knife.

Preventing discoloration

To prevent discoloration, simply leave the stone close to the avocado. Remove just before serving.

REMOVING THE SKIN

Hold the avocado gently in one hand and, with the other, peel back the skin.

AVOCADO QUARTERS

Step 1

First cut the avocado in half by placing it on the chopping board and gently cutting horizontally through the flesh towards the stone. Leaving the knife submerged in the avocado, roll the fruit around 360° to create a cut around the entire avocado. The knife should be pressing against the stone.

AVOCADO WEDGES

Place the avocado quarters curved-side down onto your chopping board and slice in half to create wedges. Be careful not to put too much pressure onto the avocado – it will damage easily.

Step 2

Remove the knife from the avocado and put to one side. Twist the avocado halves in opposite directions to separate from the stone.

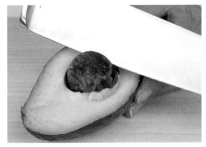

Step 3

The stone will naturally fall into one of the halves. Depending on ripeness, you may be able to remove the stone with your fingers. If not, use a knife to gently lever the stone out, trying not to cut the flesh in the process.

Step 4

To cut the halves into quarters, cut through the flesh side first rather than the skin, otherwise the skin may squash the flesh.

For avocado purée or guacamole, follow up to the end of step 3. Take the avocado halves and using a dessertspoon, scoop out the insides. Purée or cream as directed.

SLICING

Quarter an avocado then remove the skin. Take each quarter and place on one of its flat sides. With a medium or large cook's knife, slice downwards into the desired thickness. If the avocado sticks to the knife, gently slide it off (using a hollowed-edge blade will prevent this).

DICING

Step 1

A delicious way to eat avocado is by chopping it into salsa-sized dice and mixing with tomato concasse, basil julienne, olive oil and balsamic vinegar.

To dice, first slice the avocado into slices. Cut each of these pieces into 1 cm (⅖ in.) strips.

Step 2

Now cut these strips into dice. To avoid bruising, it may be easier to move the chopping board 90° instead of the avocado itself.

Mango

Mango is a wonderful fruit with many different varieties, although it can be tricky to prepare – it takes an experienced eye to predict where the large stone starts and finishes.

Think of a mango as a large almond with two cheeks of flesh attached to it. For perfect small dice of mango or mango slices, peel the mango first then remove each 'cheek', or remove the cheek first and peel with a knife – both methods are shown here.

Step 1
Steady the mango on your board on one of its ends, with the other pointing up towards you. Place the knife just slightly away from the centre and cut downwards, as straight as possible to remove one of the mango 'cheeks'. Do the same to the other cheek.

DICING

Follow steps 1–4 above, then take a knife and slice off the chunks for prepared mango dice.

WHICH KNIFE?

General preparation

A serrated knife is ideal for tackling a mango – especially if it is very ripe. For slightly more control after it has been peeled, use a medium cook's knife or Santoku knife. These knives are also great for cutting the perfect salsa-sized dice.

Cutting tip

If you slice down and come into contact with the stone, simply go back and start again, adjusting the knife's position. There will inevitably be flesh remaining on the stone – shave this off with a knife, or, as a cook's perk, eat the flesh left on the stone.

Peeling with a peeler

Much more economical than peeling with a knife, this method is best suited to a firm mango. Simply hold the mango in one hand and with a peeler in the other, peel until all the skin has been removed.

Step 2

Take one of the 'cheeks' and, using the tip of your knife, cut into the flesh. Make incisions that start at the top to the bottom of the mango and continue all the way across at regular intervals. Be careful not to score through the mango skin.

Step 3

Turn the mango around and make incisions across the width of the mango, to create a criss-cross effect. Repeat steps 1–3 for the other mango cheek.

Step 4

Now take the scored cheeks and push upwards to pop them out so that the flesh is exposed. This is a fun way of preparing mango, which can be served at this stage. If you require dice, continue to the dicing step.

PEELING WITH A KNIFE

Step 1

If the mangoes are very ripe, the easiest and most effective way to peel a mango is to cut the skin off with a serrated fruit knife or a medium cook's knife.

Prepare the mango cheeks as shown in step 1 above, and place flat-side down on the chopping board. Begin to shave off the skin with a knife, leaving as much flesh on the mango as you can.

Step 2

Continue shaving, following the curve of the mango, until all the skin has been removed. You are now ready to slice or dice as you wish.

SLICING MANGO

Take a peeled cheek of mango. With a serrated fruit knife or a cook's knife, slice downwards into equal slices of the desired width.

Pineapple

Knowing how to prepare a pineapple means you can enjoy this delicious fruit as a dessert or as part of a main course.

Ranging from a tiny 7oz. (200g) size to large 5½lb. (2.5kg) fruits, pineapples are available all year round. To test for ripeness, pull a leaf from the top of the pineapple: it should come out easily. The tastiest pineapples are picked when nearly ripe, so when purchasing, the fruit should have a sweet fragrance, without a trace of sharpness, and be heavy for its size.

Step 1
Use a serrated bread knife to remove the crown of the pineapple. Hold the pineapple steady with one hand while using the knife in a sawing motion.

WHICH KNIFE?
General preparation

Use a serrated bread knife for removing the crown and base, and for making bigger cuts into the pineapple body. For slicing smaller pieces of pineapple, a serrated fruit knife will work well.

Removing the rind

For cutting the rind off the pineapple, use a serrated fruit knife.

Pineapple rings

To make pineapple rings, use the small, serrated knife to carve out the core of the pineapple then place the pineapple on its side and cut into slices.

Step 5
Cut shallow incisions on either side of the eyes to make 'grooves' to remove the eyes. Run the grooves in a slant across the fruit's surface. Pull the 'groove' out to remove the eyes.

Step 2
Remove the base of the pineapple in the same way.

Step 3
Place the pineapple right-side up so that it is standing on a secure, flat base. Use a serrated fruit knife to remove the skin from the outside, cutting from top to bottom using a sawing motion. Notice how the 'eyes' of the rind remain.

Step 4
Cut the pineapple in half from top to bottom using the bread knife. Place the pineapple cut-side down and cut each half into half again, working from top to bottom.

Step 6
Cut the halves along their length then turn each quarter wedge of pineapple on its side and use the fruit knife to remove the core from each quarter.

Step 7
Using the fruit knife, cut across each wedge to divide the pineapple into fan-shaped slices.

Step 8
Alternatively, cut each quarter lengthways into three, then across, to produce bite-sized chunks.

Lemons and limes

Wedges or slices of lemon and limes are often used to lift and enlighten sweet and savoury dishes, calm down fiery dishes, or as a flavour-enhancing accompaniment to a number of beverages.

Lemons are often served with fresh fish as the citrus juice enhances the fish's flavour. From the same family, the smaller, rounder lime – although not as widely used as the lemon – can be prepared in much the same way. Before preparing either fruit, roll on a board with the length of one hand, back and forth – this will release the juice slightly, thereby making it easier to squeeze.

General preparation

A serrated fruit knife is the ideal choice when cutting through the tough skin of a lemon or lime (and through the juicy interior). A medium-size cook's knife or Santoku knife will also work well.

Citrus cheeks

Another way of preparing a lemon or lime wedge is to cut a 'cheek' off the side of the fruit, as shown.

ZESTING

To produce the most flavour, always zest with a knife, then cut these pieces into the desired thickness. Steady the fruit on the board and with a fruit knife, take a very thin shaving of zest. Try not to take any of the white, bitter-tasting pith. If you do, simply shave it off with a knife. Cut further into thin julienne if required. You can also zest using a peeler; this will guarantee a 'clean sweep' of zest.

CUTTING INTO WEDGES

Step 1
Top and tail the lemon (or lime) by removing a little from each end. Steadying with the claw grip, cut the lemon (or lime) in half lengthways.

SLICING

Step 1
Place the fruit on the chopping board and steady it with one hand, using the claw method (see page 32). Place the serrated knife at one end.

Step 2
Gripping the fruit, slice downwards to remove the end of the lime.

Step 3
Carve down into round slices, as thinly as desired. Continue along the length of the fruit. Aim for a uniform slice.

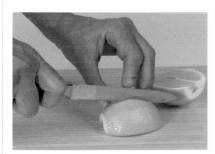

Step 2
Place the two halves flat-side down on the chopping board. Cut down the centre of each half, to form quarters.

Step 3
You may want to keep the wedges as quarters. For smaller wedges, cut each quarter into eighths, using your knife to cut on the diagonal, as shown.

Step 4
Take each wedge and, keeping it on one flat side, slice across to remove the white pith core.

Oranges and grapefruit

Similar in shape and form, oranges and grapefruits are extremely popular. Mastering the skills needed to prepare them will help you make the most of these tasty citrus fruits.

Preparing pith-free orange segments is quite a chore, but the juicy results are worth the trouble. Be careful to follow the shape of the orange closely throughout – maintaining the shape of the fruit is important to preparing a good segment.

Eating a perfectly portioned grapefruit half is a fantastic way to start the day. The method shown here enables you to portion this fruit without leaving any flesh behind, and makes it easy to eat.

PEELING/SEGMENTING AN ORANGE

Step 1
First, remove all of the skin and the pith. To create a flat base, top and tail the orange by slicing a small amount off both the top and bottom – just enough to reveal the orange flesh underneath.

PORTIONING A GRAPEFRUIT

Step 1
Position the fruit with the core and base running horizontally and, using your serrated bread knife and the claw method to grip (see page 32), cut the grapefruit in half. Saw vertically down the centre to create two equal pieces. Place the two halves cut-side up on your chopping board.

WHICH KNIFE?

Oranges

For peeling and segmenting oranges, use a serrated fruit knife. For general slicing, use a cook's or Santoku knife.

Grapefruit

To cut a grapefruit in half, use a large serrated knife, such as a bread knife. A grapefruit knife is ideal for cutting around the curves when portioning, but a small kitchen knife will also get the job done – you'll just have to work a little harder!

Orange slices

A simpler preparation method is to slice an orange. After removing all of the skin and pith (see above), slice widthways across the fruit for delicious orange slices.

Step 2
Stand the orange up on one end and slice around the edge of the orange with a serrated knife. Take just enough skin and pith off to reveal the juicy orange flesh. Turn the orange slightly and repeat down another membrane.

Step 3
Continue working your way around the orange until all the pith and skin have been removed. At this point you can slice widthways across the orange for orange slices.

Step 4
Now start segmenting. Pick the orange up in one hand, and take the knife with the other hand. Cut down along each of the segment lines, towards the centre of the fruit. Try to shave within the segments to produce the juiciest pieces possible. Repeat until you have removed all of the segments.

Step 2
Grip the grapefruit in between the thumb and first two fingers. Using a grapefruit knife (or a small kitchen knife), cut all the way around the edge of the grapefruit, between the pith and the flesh. Release as much flesh as possible from the grapefruit, leaving behind as much of the pith as you can.

Step 3
Working from the centre of the grapefruit out, cut along the segment (or pith) lines to release the segments. Again, release as much flesh as possible.

Step 4
The grapefruit segments are now ready to be lifted out and enjoyed.

Apples

A peeled and cored apple is a versatile and satisfying fruit, ideal for baking in pies, preparing a tarte tatin or simply for eating in your hand.

Peeling an apple is a practised knife skill. The aim is to remove as little flesh as possible while peeling an even amount of skin. You can use a peeler if you prefer, but using a very sharp knife gives a more refined result. Coring is another useful technique to master. When preparing apples, work quickly – like pears, apples will soon oxidize and turn brown.

PEELING WITH A KNIFE

Step 1
Grip the apple in one hand and, starting at the top of the apple, make a very shallow cut. Start to peel, following the curve of the apple. Try to keep as much of the natural shape as possible.

WHICH KNIFE?

Peeling

A small kitchen knife is the most comfortable to use for this technique. Obviously, a peeler can also be used.

Coring

Use a small kitchen or turning knife. Corers are available but can be restricting when the core isn't straight (i.e. sometimes in cooking apples).

Coring variation

For a more organic finish to a cored, quartered apple, cut the core out in a curved shape. This is more pleasing to the eye as more of the apple shape remains. Hold an uncored apple quarter in one hand and with either a small kitchen knife or a turning knife simply start carving at one end. Gently follow the natural shape of the apple using an up-and-down motion.

Step 3
Stand a quarter piece on its flat side and cut the core out horizontally, taking the least amount of apple flesh away as possible. Repeat to all quarters.

Step 2
Once you need to adjust the apple in order to continue cutting, simply stop cutting and, without moving the knife, shift your hand around the apple until you can resume cutting. Continue cutting in a spiral motion until you reach the bottom of the apple.

APPLE GARNISH

Step 1
For a three-tiered apple garnish, cut apple halves into a series of 'V' shapes. Lay an unpeeled apple half flat on a chopping board. Take a medium cook's knife and cut into the apple around ⅕ in. (5 mm) from the bottom, cutting downwards at an angle, towards the centre, as steadily as possible. Try not to move any part of the apple.

CORING

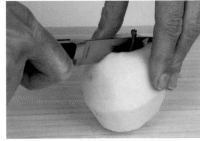

Step 1
Place a peeled apple on a chopping board and cut in half using the under-the-bridge method of cutting (see page 30). Slice downwards, right through to the board, using a small kitchen knife.

Step 2
Once you reach the centre, mirror the cut on the other side of the apple. Next, make the same cut 1 cm (⅖ in.) higher, through to the centre, then mirror again on the other side. Repeat the process to create three or four layers.

Step 2
Place each half flat-side down on the chopping board and using the claw method to grip (see page 32), slice into quarters.

Step 3
Once you have finished cutting your layers, gently push them apart to make an attractive garnish. Aim for equal contrasting rows of colour.

Pears

Pears can vary enormously in the variety of texture and consistency – a pear can go from under-ripe to very soft overnight. For a soft pear, a sharp peeler is best equipped to give uniform peeling results.

Coring is the ideal method of preparing a pear for poaching. For a clean presentation, remove the core using a melon baller (see page 19).

Preparing pear halves, as shown opposite, is an essential skill for baking upside-down pear cake or tarte tatin.

PEELING

Holding the pear with a firm grip in one hand, use your peeler to peel the skin from the pear. For the smoothest possible finish, especially if the flesh is soft, guide the peeler along gently.

WHICH KNIFE?

Peeling

A sharp kitchen knife or turning knife is good but with a very soft pear, these knives can slip easily. A sharp peeler is best suited to peel uniformly.

Coring

Use a melon baller for coring a whole pear or pear half. To core a quarter, use a serrated fruit knife or kitchen knife.

Coring a quartered pear

In order to quarter a pear, which is much like deseeding a tomato or coring an apple, place the quarter on its side and take out the centre core in a diagonal cut, as shown below. You can also follow the natural line of the pear by holding the quarter in your hand and cutting an organic shape around the core (see *Coring Variation*, page 102). This method should only be practised by more experienced knife users.

Step 2
Scoop out approximately half a ball of pear, removing the entire core without taking too much of the pear flesh in the process.

CORING A WHOLE PEAR

Step 1
Place a peeled pear on its side and slice a little off the bottom (this ensures that the pear can stand up nicely during serving). Only slice off a little, so that the majority of the fruit remains.

Step 2
The centre of the core is identifiable by a brown dot located in the middle of the pear. Place the melon baller in the center, pierce the flesh and scoop around using a 360° motion. Try to remove as large a ball as possible from the bottom of the pear.

Step 3
Remove the core line where the stalk is. Using a serrated fruit knife, or a kitchen knife, dig out the core line on either side of the stalk, in a 'V' shape, again making sure to leave as little of the core line behind as possible.

Step 4
Pull out and remove the stalk – it should come out easily in one step.

PREPARING PEAR HALVES

Step 1
Cut a whole, peeled pear in half lengthways to create two equal halves. Place the halves curved-side down on a chopping board. Take a melon baller and position over the lower third of the pear where the majority of the core is located. Pierce the flesh.

Step 5
Remove the bottom end of the core in the same way as the top end, by cutting a small 'V' shape.

Stoned fruit

The knife skills demonstrated here can be applied to all stoned fruit – nectarines, peaches, plums and more – although they can be difficult to implement if the fruit you are working with is under-ripe.

It is important to take care when halving a stoned fruit, as the knife can be damaged on the hard stone. To aid the peeling of peaches or nectarines, pour over boiling water and let them sit in it for about 15 seconds, before removing. Use a turning knife or a kitchen knife to get under the skin to remove.

WHICH KNIFE?
General preparation

A serrated fruit knife is universally good for the preparation of all stoned fruit. For slicing super thinly, use your favourite cook's knife or Santoku knife.   For small dice, try using a kitchen knife. In all cases handle knives with care when cutting around the stones of these fruit.

Stoning tip

The stone can be impossible to remove when fruit is under-ripe. If this is so and your recipe allows, simply leave the stone in until cooking finishes. Alternatively, cut the stone away with a serrated fruit knife, sticking as closely as you can to the stone.

HALVING AND STONING

Step 1
Steady the fruit on your chopping board using the claw method (see page 32). Position your serrated fruit knife and cut through the centre.

FOR WEDGES OR QUARTERS

Step 1
Follow steps 1–4 above. Turn the halved fruit onto its flat side, and using the claw method to grip, cut in half as equally as possible.

Step 2
When the knife hits the stone, gently roll the fruit, keeping the knife in one place while cutting around its equator. Continue until you have cut around the entire circumference of the fruit.

Step 3
Put the knife down, take the fruit in both hands and twist in opposite directions. The fruit should become loose at this stage.

Step 4
Gently pull the two pieces apart. The stone will remain in one of the halves. Remove the stone by hand or use the tip of a knife to nudge the stone out. Some exceptions to stoning apply – for a tarte tatin, for example, leave the stone in for cooking, and remove before serving.

FINELY SLICING HALF-MOON PIECES

Step 2
Turn the pieces onto their curved side and slice into the chosen size of wedge – either quarters, thirds or even smaller wedges.

Step 1
Follow the halving and stoning steps above. Then, place a fruit half flat-side down on your chopping board. Grip with the fingers of one hand, and take a serrated fruit knife in the other hand.

Step 2
Slice as thinly or thickly as you wish, at regular intervals, all the way across the fruit.

meat and

poultry

Here you'll find basic, intermediate and a few advanced butchery skills needed for preparing the most common types of meat. Whether you are slicing chorizo, mincing beef, butterflying a leg of lamb or jointing an entire chicken, this section shows you the easiest and quickest way to do so.

Meat and poultry knives

Good knife skills are an important ingredient in meat preparation. Knowing the right knife to use for each task, and how to use it, makes cooking easier, faster and safer. The best meat knives have blades forged from high-carbon stainless steel. The most durable will have rivets through the handle to hold the blade in place.

Other useful tools

POULTRY SHEARS
Arguably two knives in one, poultry shears are an incredibly sharp, robust pair of scissors. Ideal in all poultry preparation, they will cut through most small bones (see page 19).

MEZZALUNA
The curved double blades of the mezzaluna make it a great tool for mincing your own meat but two equally-sized cook's knives can be used for the same task (see pages 120–121) .

Cook's knife/Santoku knife
The cook's knife is the most frequently used knife in the kitchen. It comes in sizes ranging through 15–30 cm (6–12 in.). It is essential for chopping, slicing and mincing. When preparing meat, it is mainly used for slicing and mincing: two cook's knives of the same size can be used together to mince meat by hand.

Boning knife
For all boning and butchery, the boning knife is ideal. It is designed to be used in a dagger-like fashion, with the tip being the most important part of the knife (see page 35). The rigidness of the blade encourages precise and definite cuts.

Carving knife

The carving knife is essential if you are to carve your cooked meat effectively. It has a long, thin, ridged and slightly curved blade. The length of the blade encourages a long, sweeping slice and its hollowed edge prevents the meat from sticking to the blade. The most useful size is 25–30 cm (10–12 in.) long. A longer knife will give you more even slices and will be easier to use. Used in conjunction with a sturdy fork with long, thin tines, a carving knife should have a finger guard between the blade and the handle to protect your fingers in case your knife slips.

Butcher's knife

The butcher's knife has a sturdy blade, approximately 15 cm (6 in.) long. It has a narrow blade with a flange next to the bolster, which helps keep the knife from slipping into the meat. It is often held with a 'stabbing' grip when being used to bone large joints of meat.

Flexible knife

The flexible knife is also known as a fish knife as it is used most frequently for filleting and skinning fish. It has a very thin blade, usually 20 cm (8 in.) long. The flexible blade is useful for removing the membrane from tenderloin cuts and removing sinew from all cuts of meat.

Meat cleaver

The large, rectangular blade of this knife is designed for heavy jobs and to cut through bones – in fact it is one of the only knives that should be used to cut through bones. A regular knife will be damaged if used for this purpose.

Chicken, turkey and duck

Poultry is a delicious and versatile food group that calls for a number of preparation techniques, depending upon your preferred way of eating it.

Whatever your choice of poultry meat, learning to joint a bird – separating the bird at its joints – is a great knife skill to learn. It is much cheaper to joint a bird and make use of its parts, than to purchase these separately. Instead of being limited to various cuts, jointing a bird opens up a world of recipes, allowing you to use the carcass of the bird to make stock, or to render the fat from the skin for roasting potatoes.

JOINTING A CHICKEN OR TURKEY

Step 1
Place the bird on a chopping board and untruss by cutting through any string or elastic.

Step 5
Remove the legs. First grip the skin located in the middle of the breasts. Pull upwards with your fingers – this ensures the even distribution of the skin at a later stage. Then, where the legs meet the main body of the chicken, use a knife to cut just through the skin without piercing the flesh. The leg will fall to one side. Repeat for the other leg.

WHICH KNIFE?
Jointing/filleting/slicing

Poultry shears are very useful for the preparation of this type of meat and are perfect for cutting through small bones with ease.

A rigid boning knife is great for filleting, while a medium cook's knife is ideal for slicing and cutting through joints, as is a meat cleaver or bone splitter (see page 16).

To remove the oysters from the bird, use a kitchen knife.

The oysters

Never forget to remove – and eat – the oysters from a roasted chicken. Turn the chicken over and look for two oyster-shaped nuggets of succulent meat located at the top of the chicken legs. Having been constantly basted with chicken juices during cooking, these discs are very juicy and tasty.

Step 2
Turn the chicken breast-side down and remove the rump of the chicken, or the pope's nose. To do this, take a cook's knife and make a clean cut, close to the carcass. You may want to retain the pope's nose as it is rich in fat and flavour, and depending on your taste, can be one of the highlights of a roast chicken.

Step 3
Keeping the chicken breast-side down, cut through the skin in the centre of the chicken (along the backbone). Cut from the top to the bottom. This makes removing the legs easier later on. Note that the knife doesn't travel very far down at this stage – just through the skin and it will hit the bone.

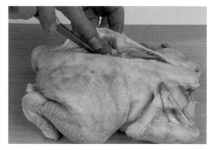

Step 4
Now release the oysters located at the top of the legs. Follow the lines of the thighs towards the backbone and there they lie. Cut the flesh away from the sockets with a small kitchen knife. Keep the knife close to the carcass and scrape to remove both oysters. Turn the chicken over, breast-side up.

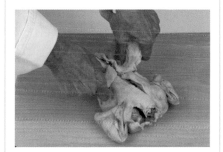

Step 6
Pop both legs out of their sockets. This will make the chicken easier to handle.

Step 7
Hold the chicken by one of the legs and let the carcass hang down. Using your knife close to the carcass, join up the cuts made in steps 4 and 5, to remove the leg. You shouldn't be cutting through any bone, just flesh and skin. Repeat on both sides. You should be left with two leg pieces and the main carcass (the crown).

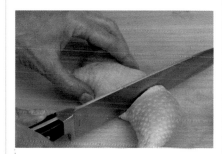

Step 8
Divide each leg into a drumstick and thigh. First locate the ball and socket joint that connects these two pieces and make a clean cut through the joint. You shouldn't have to cut through any bone – the knife can glide straight through the ball and socket joint. If the knife comes into contact with any bone and resists, take it out and reposition.

Continued next page

Step 9
To remove the backbone, use poultry shears to cut along both sides of the bone – the poultry shears will cut through any small bones fairly easily. Then, remove the backbone and either discard or, alternatively, use for stock.

Step 10
Place the crown of the chicken flat on the board. Cut through the centre of the breastbone – the knife will naturally fall to one side of the bone – from the top to the bottom of the crown. Cut straight down, keeping your knife close to the bone. Pull the flesh away slightly so that the breastbone becomes visible.

Step 11
Using the poultry shears, cut along the scored side through the breastbone from the bottom to the top of the crown. There will now be two pieces.

REMOVING THE BREAST HALVES

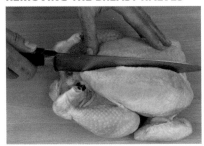

Step 1
Place the bird on a chopping board and, using a cook's knife, cut through the breast. Let the knife fall naturally to one side of the breastbone.

Step 2
With your knife in one hand, gently cut the flesh away from the carcass, pulling the breast half back with the other hand. Use definite, sweeping motions while you cut and make sure that your knife is in contact with the carcass at all times – this will ensure as much flesh comes away as possible.

Tip
When removing the breast from a whole bird, let the breastbone be your guide. Keep your knife as close to the breastbone as possible throughout the process.

Hygiene
When preparing raw chicken or turkey, be especially careful of the the natural bacteria that resides in the raw meat. Wash all knives, surfaces, chopping boards and hands well in hot soapy water.

Step 12

Now separate the breasts into four portions. Keeping in mind that a chicken breast is thicker at the top, use your cook's knife to cut downwards on a diagonal and divide the breast into two equal portions. Cut until your knife hits the breastbone, then remove the knife so that the meat is only partially cut.

Step 13

Take the poultry shears and cut through the bone completely. Repeat with the other breast.

Step 14

There should now be eight portions of chicken. Note: the drumsticks can be trimmed further at the red lines.

Step 3

Once the breastbone is off on one side, cut the breast off entirely with poultry shears, or using your cook's knife. Repeat with the other side.

SKINNING A BREAST

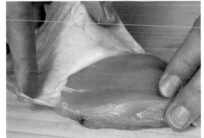

Place the bird on your chopping board. Hold the breast under the skin at the thicker end and, with your free hand, pull the skin in the opposite direction (away from your other hand). Remove the skin fully.

SLICING THE BREAST

Place the breast on the chopping board. At a 90° angle to the direction of the grain, slice downwards into pieces of the desired width.

Step 1

Remove the chicken from the roasting tin, draining and retaining any delicious juices. First remove the legs by steadying the bird with a carving fork and cutting down between the leg and the carcass. You do not have to cut through any bone – the leg should pop out easily. Repeat with the other side.

Step 2

Divide the legs into drumsticks and thighs as you would with a raw bird (see step 8, page 113).

Step 3

Carve the breast into slices as thin as you like, taking the delicious skin with you. Continue until most of the flesh is removed from both breasts. Serve the meat immediately.

SPATCHCOCKING

Carving tips

Before carving it is important to let the cooked bird rest (as with any meat). This allows the meat time to 'set' and ensures it is firm enough to carve properly. Specially designed chopping boards with space to collect the juices are available for this purpose. Don't throw out the juices that collect, always add them back to the gravy.

Ideally use a special carving fork and knife for carving, but if you don't have these, try using a cook's knife and eating fork instead.

Spatchcocking

Spatchcocking poultry is a great method to master as it enables you to cook a bird on the barbecue. Spatchcocking basically entails cutting out the backbone of the bird and flattening it out. The legs are flattened in an inward direction and skewers are inserted to keep the position of the bird in place. You can also use this method with poultry shears to prepare poussin or quail, or indeed any feathered game.

Step 1

First remove any trussing string and turn the bird breast-side down. Cut down both sides of the backbone with poultry shears before removing the bone. Use for stock, or discard.

SCORING A DUCK BREAST

Using your cook's knife, score through the skin to create diamond openings, but do not pierce the flesh – due to the soft nature of the skin, this is easily done, so take care.

SHREDDING DUCK

Step 1
Place your slow-roasted crispy duck on your chopping board. Using two forks, pick off chunks of the meat and skin.

Step 2
Continue pulling the meat apart using the two forks. Work along the natural lines of the cooked meat and keep going over the meat until it is well shredded. Use this technique just prior to serving – the meat will lose heat quickly once it has been shredded.

Step 2
Turn the bird breast-side up then turn the legs inwards. Using the heel of your hand, press down firmly in the centre of the breast. A slight crunching sound indicates that the bones and the breastbone are breaking.

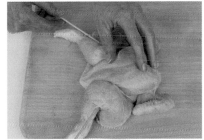

Step 3
Carefully insert a skewer just under the surface of the bird, and through the leg.

Step 4
Insert a second skewer up through the breast, so that the skewers form a cross shape. You are aiming to create a flat surface in order to ensure even cooking.

Beef and veal

Good beef should be a deep red colour with a solid creamy layer of fat and plenty of marbling (fat running through the meat). Veal is beef that is no more than a year old. With its characteristic pale pink flesh, this young, tender meat has a delicious and mild flavour, and can be prepared in much the same way as more mature beef.

The tenderloin is the most tender cut of beef, and although not the most flavoursome, it's definitely one of the most superior (and expensive) cuts. As with most meats, it is cheaper to buy a whole tenderloin and prepare it yourself. Individual pieces can be frozen for later use.

WHICH KNIFE?
Cutting/removing sinew

A rigid boning knife is ideal for trimming beef and removing sinew – the thin tip of this knife can easily get under the sinew and run along it.

Use a large cook's knife or classic butcher's knife to cut beef steaks.

For cutting veal into escalopes, use a long carving knife, such as a slicing knife or a ham knife.

WHICH KNIFE?
Carving

To carve a joint, a classic carving knife and fork are ideal. Alternatively, use an eating fork and the longest cook's knife you have.

Mincing

For mincing by hand, two identically sized knives (large cook's knives ideally) should be used. If you have access to it, use a mezzaluna.

PREPARING A TENDERLOIN

Step 1
Using a boning knife, or a cook's knife, trim off any excess fat and sinew from the fillet. This is known as removing the 'false fillet'. You can cut against the grain into thin strips for a stir-fry if you like.

PORTIONING A TENDERLOIN

Step 1
Grip the beef in one hand and slice through with a large cook's knife. Use one cut moving in a forward motion, to separate a one-third portion. If the fillet is trimmed properly, a perfectly sharp knife will glide through the meat, much like cutting through butter. This third is called the chateaubriand (the thicker end of the fillet).

Step 2

Once you've trimmed the excess sinew away, trim any remaining sinew. Sinew is identifiable by its shiny, silver colour. Start at one end of the tenderloin, insert a cook's knife (or boning knife) just underneath the sinew and cut along the length.

Step 3

Grip the released sinew and pull it tight while continuing to cut along underneath it. Keep your knife at a slight angle so as not to cut through the sinew – if you do, simply replace your knife underneath the sinew and continue cutting.

Step 4

Trim away the fat from the tenderloin, as required. You may want to retain a little fat – it will provide flavour and moisture during the cooking.

Step 2

Cut the remaining two thirds into halves, so you have three thirds.

Step 3

To cut into tournedos, slice the meat into pieces approximately 2.5 cm (1 in.) in width. For medallions, cut into even thinner slices. Thin slices for stir-fries or stroganoff can also be cut from here.

Preparing carpaccio

The word 'carpaccio' is now a loose term used to describe something raw and essentially thinly sliced.

Many cuts of beef are suitable for carpaccio, including tenderloin, sirloin and rump. The only rules are that beef should be cut evenly and thinly. If you have access to a meat slicer (see page 21), use it. If you are using a knife, freeze the beef first for 20–30 minutes to steady the meat before you begin cutting. Rolling the beef tightly in cling film to create an even round shape prior to cutting is also helpful. (See *Tuna Carpaccio*, page 153).

MINCING

Step 1
This technique can be applied to any type of meat. Place your chosen cut(s) of beef on a chopping board and slice into evenly sized pieces.

Step 2
Gather the slices together, turn 90° and cut into evenly sized dice as shown.

Step 3
With one hand, take two cook's knives and grip the handles together. Lean the tips of the knives against the chopping board and start to cross-chop (see page 34). Keep the tips of the knives in one place and pivot the handles back and forth across the meat pieces.

Mincing

Mincing breaks down meat fibres and makes tougher cuts much more digestible. This is ideal if you want to blend your own burger mix — simply purchase individual cuts and mince them together. Whatever your chosen meat ingredient, always ensure it is as evenly chopped as possible. Be aware of hygiene too — remember, the more cut sides an ingredient has, the higher the risk of germs.

Tip

Mincing can be quite a lengthy process, and during this time the meat can get warm. If it does, simply place the meat in the fridge until it has cooled. Repeat as required.

CUTTING VEAL ESCALOPES

Cutting your own escalopes, like mincing your own meat, is often reassuring for the eater. The ideal cut for an escalope is top round. For the best results, examine the meat first to see where the grain lies, then very thinly cut across at a 90° angle to the grain, as shown.

Step 4
Continue chopping until the meat is finely minced. While chopping up and down, stop occasionally and scrape the meat together with the knife, therefore preventing the meat from dispersing across the chopping board too much.

CARVING

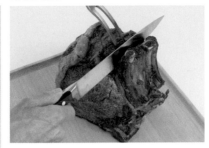

Step 1
When the meat is fully rested, place on a chopping board on its flattest side and steady with a carving fork. Begin removing the bones by placing the knife under and very close to the bones. Maintain contact between the carving knife's blade and the bones.

Step 2
Allow the knife to glide down the bones at a slight angle, leaving as much flesh behind as possible. Cut until the bones are removed.

Step 3
Once the bones have been removed, start to carve the main body of the meat. Keep the carving fork in place to steady the beef. Carve the meat in a carving motion – literally moving the knife backwards and forwards. Carve as thin or as thick as you like.

Step 4
If the meat is too rare in the middle (sometimes this happens with very large cuts), start to carve from the other end, using the technique detailed in step 3. Meat is often better rare than overcooked. When you get to the centre of the roast, it may be easier to halve again before continuing to carve.

SLICING A SIRLOIN STEAK

Use a fork or a pair of tongs to hold the steak in place. Using a carving knife or a regular cook's knife, slice the steak on the slant – at a 90° angle to the grain – as thinly as desired.

Lamb

Lamb is a dark, red-colored meat and should have a good layer of fat. Like any meat, lamb is available in various cuts. A leg of lamb is a great cut to use for butterflying because it is less fatty than the shoulder. Although fattier, the shoulder is, however, much sweeter than the leg.

Butterflying a leg is a key knife skill to learn. The technique makes the meat more manageable than cooking a whole leg on the bone. It also provides the option of cutting the leg into steaks or dicing the meat for a casserole. While butterflying, it's helpful to visualize or feel where the bone lies.

WHICH KNIFE?
Boning and carving

The rigid boning knife is essential here. For boning techniques such as butterflying a leg or tunnel boning, use the knife in a dagger-like motion.

When carving, a carving knife is ideal, but you can use a large cook's knife for the task.

For all other preparation techniques, use your favourite cook's knife. For more control, use a small kitchen knife for intricate work such as trimming a rack of lamb.

The boning knife

The boning knife can be used in the traditional way when removing sinew, however, when tunnel boning a shoulder or butterflying a leg, use the knife in a dagger-like motion, as shown on page 35. The tip of this knife is designed to come into contact with the bone and remove the flesh from it.

BUTTERFLYING A LEG

Step 1
Start to remove the bone by releasing some of the flesh from it. Use the boning knife to cut away the flesh from the bone, letting the tip of the knife slide next to and press directly against the bone.

Step 5
Having cut it away as efficiently as you can, lift the bone out. Once the bone is removed, use to make stock or discard.

Step 2
Start to open up the thickest part of the leg. Bear in mind the shape of the bone you are removing and where possible, let it guide you.

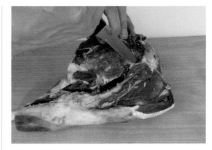

Step 3
Use definite, confident movements to dig the bone out. Release as much of the flesh from the bone as possible.

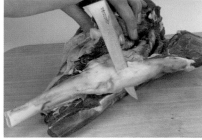

Step 4
Continue to cut behind the bone and around the joint.

Step 6
Now concentrate on butterflying, or flattening out, the remaining piece of meat. Simply use a knife to make long, shallow incisions into the flesh and flatten out the meat. This can also be carried out with a meat tenderizer or a rolling pin covered with cling film.

Step 7
Remove any sinew and fat that you can see, using the boning knife, this time in a traditional knife-holding way.

Step 8
The result should be an even piece of meat, ready to marinate and throw on the barbecue.

CUTTING A LEG INTO STEAKS

Step 1
Taking a boned leg, select a portion that is thick enough to be cut into decent steaks (i.e. not too thin, so the leg steaks can be cooked rare). Cut the portion across the grain (the natural lines in the meat), using a cook's knife.

Step 2
Cut each steak to a thickness of at least 3 cm (1⅕ in.) – especially when preparing to cook steaks medium rare.

Step 3
Continue cutting. Try to produce evenly sized pieces. Fold the meat around to create a neater appearance if you wish. You may want to trim all sinew and fat at this point, as the sinew can shrink during cooking and distort the result.

CARVING A COOKED LEG OF LAMB

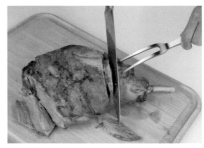

Step 1
Steady the cooked, rested leg on a chopping board with a carving fork. Using a carving knife (or a large cook's knife) start carving at the far end of the leg. Carve as much as you can, and as thinly as you like, until you reach the bone.

Step 2
Proceed further down the leg, until you reach a particularly fleshy part. Cut a 'V' shape – almost a wedge of meat – into the leg (this is the only way to be able to cut across the grain). Use the belly of the knife at this point. Cut down towards the leg, on the diagonal.

Step 3
Repeat the cut on the other side, aiming for a "V"-shaped piece of meat. Remove the cut piece and place on a board. Slice the piece further if desired. Continue to carve the leg using this technique.

DICING THE LEG

Step 1
Remove a clean piece of meat from the leg. Start to divide it according to its natural partitions, using a cook's knife.

Step 2
Trim away as much sinew as possible – although fat equals flavour, so retain as much fat as desired. This knife skill is rather like that of skinning a fish.

Step 3
Leave the skin touching the surface of the chopping board, taking away as much flesh as possible.

Tip

Cooking a leg on the bone retains more of the meat's flavor and moisture. However, this does make carving tricky as the bone is large and often gets in the way. Follow the steps opposite for perfect carving. Always carve at the last minute to retain heat and moisture.

Step 4
Trim any further pieces of fat or sinew. When the meat is 'clean' of sinew, start to cut it into evenly sized strips.

Step 5
Turn each of the meat strips 90° and slice further into dice. Try to keep a square shape wherever possible, and don't cut too small: 5 cm (2½ in.) squared is ideal.

TUNNEL BONING A SHOULDER

Step 1
Place the shoulder flat on to the chopping board, skin-side down, with the natural opening placed in the uppermost position.

Step 2
Hold the boning knife like a dagger (see page 35) and insert it deep into the widest part of the shoulder, keeping close to the shoulder blade. With the tip of the knife next to the shoulder blade, make definite sweeping cuts to release the flesh from bone.

Step 3
Next, turn the shoulder over to cut away flesh from the other side of the bone, then turn the shoulder back over.

Step 7
When you have cut as far as possible, start tunnelling from the other end of the bone. Score around the bone at the base using the tip of the knife.

Step 8
Hold the bone up and insert the knife where the flesh joins the bone. Loosen the flesh surrounding the bone, being careful not to cut through any of it. Separate all of the flesh that connects to the bone. When all of the flesh has been fully detached from the bone, the bone can be carefully removed.

Step 9
Use your hands to remove the bone. To do this, you may have to release the bone from very small connections still. Use your knife to do this.

Step 4
The shape of the shoulder blade should become more and more apparent. As you cut, continue to dig around it.

Step 5
When you finally reach the bottom of the shoulder blade, you will come to a ball-and-socket joint. Cut around it, opening the lamb up as little as possible. This will take time.

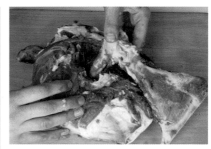

Step 6
After cutting around the ball-and-socket joint, cut a little further down into the flesh. Cutting may become more difficult as the meat clings tightly to the bone.

CARVING A SHOULDER

Step 10
The shoulder will now be boneless and ready for stuffing or rolling.

A tunnel-boned shoulder is usually stuffed or rolled, but if you decide to carve, it is easily done. The lack of bone means you can carve as you like, ideally against the grain.

To carve on the bone, as pictured here, visualize the 'N'-shaped bone in the shoulder and use 'V'-shaped wedge cuts – rather like cutting a cake – in between the bones, as shown above.

Tunnel boning

'Tunnel boning' literally involves making a tunnel in the meat, or not making any holes in the flesh apart from removing the bone. This method is ideal if you want to stuff a shoulder or stuff and roll a shoulder. The shoulder consists of a large, flat shoulder blade. Like other boning processes, keep your knife close to the shoulder blade all the time, using it to guide you.

ENGLISH TRIMMED RACK

Step 1
Place the rack on the chopping board with the 'chine' bones (backbones) facing downwards. Using a kitchen knife, remove any skin (sometimes the butcher will do this for you) but try to leave most of the flesh behind in the cleanest way.

Step 2
Remove the chine bone by cutting the flesh away – remember to let the boning knife scrape against the chine bone at all times. As more flesh is released, the rack will open up and become easier to manage. Once the chine bone is detached, discard or use for making stock.

Step 3
Remove the shoulder blade, which is recognizable as a half-moon shaped cartilage. At the thinner end of the rack, part of the shoulder blade is occasionally left behind. If this is the case, remove by simply pulling it out and then discarding.

The trim

An English trimmed rack of lamb simply means that the fat and bones are trimmed slightly for a more refined rack. When a rack of lamb is French trimmed the fat and bones are trimmed right down and a bare minimum of fat is left on the bones – leaving a very clean cut. Trimming a rack of lamb takes time to prepare, so if purchased ready trimmed, it is expensive. Mastering this technique at home makes this cut affordable. Request the best part of the neck and always ask the butcher to 'chine', or to sever, the backbone refined meat.

Step 7
Stand the rack up and start to remove the meat from in between the bones. Then scrape the bones using your knife, and rubbing up and down. This process involves lots of scraping and is time-consuming – patience is required. If any flesh is left on the bones it will burn in the oven and look unsightly.

Step 8
Try also to lie the rack down on the board and scrape. Find a position that is most comfortable for you.

Step 4
Remove the paddywhack neck tendon. Begin by placing the lamb flesh-side down on the chopping board. Look for a creamy coloured long, skinny tendon. Insert the boning knife, or a kitchen knife, just under the tendon, release a little and then pull it away.

Step 5
Position the lamb flesh-side down and fat-side uppermost. Measure from the top of the bones roughly about 3.75 cm (1½ in.), and score through the fat using your knife. Note that the knife will come into contact with the bones.

Step 6
Dig the knife underneath the fat. With the knife alongside the bones, remove the strip of fat and meat with your other hand.

FRENCH TRIMMED RACK

Step 1
Follow steps 1–4 for the English trimmed rack above, then completely remove the fat and skin from the rack.

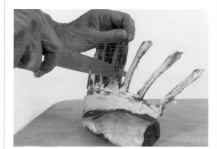

Step 2
Use your knife to clean the bones, right down to the meat, separating them as you do so. This will take time so be patient.

Step 3
Clean the main tenderloin of lamb by removing the sinew. Insert a boning knife (or cook's knife) under the sinew and cut along underneath the sinew very carefully.

Carving a cooked rack of lamb

The presence of bones makes carving a rack of lamb relatively easy – just let your knife lean on one side of the bones and cut down. Before you begin carving, assess the natural division of meat between each bone. Ensure that each bone has an even amount of flesh on it.

(see glossary pages 172–173, for definition)

CARVING A COOKED RACK

Hold the bones with one hand and make a clean, sweeping cut through the meat – let your knife lean on one side of the bones and cut down. Continue cutting until all the meat is divided.

SHAPING GUARDS OF HONOUR

So-called because the bones in this cut interact like crossed swords, two French trimmed racks of lamb (see page 129) and some butcher's string or cooking twine are all that is required to prepare this cut. Line the racks up next to each other, trying to match the ends of both together. Allow the bones to cross equally and tie the racks together.

Step 2
Cut the tenderloin away from the bones, keeping your boning knife close to the bones at all times and taking care not to cut into the eye (see glossary pages 172–173, for definition) of the meat. Discard the bones or use them to make stock.

Step 3
Trim the fillet by removing any remaining fat and sinew. Try to retain the original shape of the fillet as much as possible.

Step 4
Remove the 'false fillet' or smaller tenderloin that runs alongside the main tenderloin, and set aside for another recipe. Clean the main tenderloin of all fat, sinew and connective tissue.

SHAPING A CROWN ROAST

A celebratory centrepiece, the resemblance of a crown earned this dish its name. Stand two or three French trimmed racks up in a circular or oval shape, with their backs facing inwards. Arrange the curved shape until it is even and satisfactory and tie in place with some string or cooking twine.

Noisette: the definition

Noisette is French for 'hazlenut', and in the context of cooking lamb refers to the heart, or core, of the cut of lamb. This is yet another method of using the neck. In this instance, the bones are not used at all. The fat is trimmed off and tied around the tenderloin with string (or cooking twine) and then usually pan-fried or pan-roasted.

PREPARING NOISETTES OF LAMB

Step 1
Remove the fat from a rack of lamb by digging your boning knife under the fat and then using your fingers to pull it away carefully. Set aside to trim later.

Step 5
Trim the fat that was removed in step 1. Aim to create an even, flat, square surface that will fit the tenderloin. Then, roll the fillet up in the fat to create a perfect cylinder of lamb.

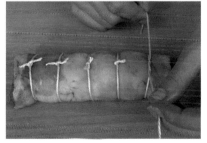

Step 6
Tie the meat up in an even series of secure knots.

Step 7
Portion these into even noisette slices, between 2.5–3.75 cm (1–1½ in.).

Pork

Pork is a fantastically versatile meat with many different preparation techniques and products, including ham, pork belly, salami and lardons.

A tenderloin cut of pork is cheap, quick to cook and tender. It marries well with many flavors and little preparation is needed to enjoy it. It can be cooked as a whole and then sliced, or sliced first (as shown here) and cooked as medallions or 'battened out' as escalopes. Use a rigid boning knife to trim, followed by a medium cook's knife. The white membrane on the tenderloin needs to be removed before cooking. It is not only tough and chewy to eat, but shrinks when the meat is cooked, and distorts the shape of the meat.

WHICH KNIFE?
General preparation

For large scale, general butchery a special butcher's knife is ideal. Otherwise a cook's knife is good for general pork preparation purposes, particularly portioning a tenderloin.

Slicing sausage

A specialist serrated sausage knife is good for slicing chorizo or other cured sausages.

WHICH KNIFE?
Removing a membrane

For trimming of fat or sinew and boning, use a rigid boning knife. The thin tip is designed to easily slip under the membrane. A flexible meat knife of the same shape can also be good here – it's a matter of personal preference.

Carving

Use a ham knife or a long, thin-bladed knife and a carving fork.

PORTIONING A TENDERLOIN

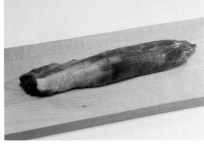

Step 1

The white membrane on the tenderloin needs to be removed before cooking. It is not only tough and chewy to eat, but shrinks when the meat is cooked and distorts the shape of the meat. Look for the shiny sinew – it covers approximately one quarter of the loin. There may also be a layer of fat.

Step 5

Once all the sinew is removed, start to trim any creamy fat. Remember, however, a little left on is great for flavour and also bastes the meat during cooking.

Step 2
To remove the sinew, hook the boning knife just under it at one end. Loosen the sinew and cut along, slightly underneath it. As soon as it is possible, grip it with your non-knife hand.

Step 3
Pull the loosened sinew taut and position your knife at a slight angle underneath it, so the sharp blade edge is tipping upwards towards the sinew. (This will ensure enough of the flesh is removed from the sinew.) Continue to cut along. As the sinew is tough, the knife shouldn't go through the sinew.

Step 4
The sinew will disappear from the surface of the meat – simply stop cutting when you get to this point. Then begin the process from step 2 again to remove any remaining sinew, turning the tenderloin over if necessary.

Step 6
When the tenderloin is clean of any fat or sinew, you can roast or sear as a whole, or cut the tenderloin into steaks of the required thickness – 1–2 cm (⅖–¾ in.) is a good thickness for sautéing the steaks.

Step 7
The tail end of the whole tenderloin tapers into a narrow point that is too thin to cut into steaks across the grain. Leave this in a long piece, cutting approximately 10 cm (4 in.) from the end of the tenderloin. Turn 90° and press out into a steak, against the grain of the meat.

Step 8
Turn the rest of the pieces so the cut edge of the meat is facing upwards for cooking. The meat will then be tenderer to eat as you will bite along the length of the grain, instead of through it. Either season and pan-fry, or beat to tenderize for quicker cooking.

SCORING A PORK BELLY

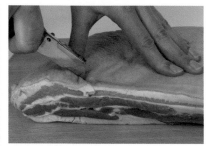

Step 1

Start at one end of the belly by inserting the tip of the utility knife to a depth of around 5 mm (⅕ in.). Hold the belly firmly in place with the fingertips of one hand.

Step 2

Pull the knife through the skin firmly, dragging backwards until you reach the other end of the belly. Aim to keep a 5 mm (⅕ in.) depth all the way along.

Step 3

Continue and repeat at regular 1 cm (⅖ in.) intervals, all the way across the width of the belly.

The pork belly

The amount of fat on a belly varies according to the breed of pig. For a more efficient rendering of the fat and for easier slicing once cooked, score the skin. This is difficult to manage with a cook's knife because the skin is so tough. The best tool is a very sharp utility knife and some strength! Otherwise ask the butcher to do this.

Lardons

These are lovely chunks of bacon cut into pieces approximately 1 cm (⅖ in.) in width. They originate from the same piece of meat as streaky bacon and pork belly.

CUTTING LARDONS

Step 1

Place the pork skin-side down on a chopping board (although you may want to trim the skin off before). Use a long carving knife, or the largest cook's knife available, and cut a slice of bacon about 1 cm (⅖ in.) thick. If the bacon is too long, simply slice in half to make pieces more manageable.

Step 2

Repeat all the way along the piece of bacon. Aim for consistent slices.

CARVING A HAM

To carve a ham on the bone, begin by cutting away any fat, then slice thinly across the grain. If it is glazed, try to distribute as much of this as possible across each slice.

BONING A COOKED HAM

It is easier to take a ham off the bone before carving, although it doesn't look as attractive. Remove the three natural lobes (hip, thigh and shank) with a carving knife or any long knife. Then transfer the ham to a chopping board and carve across the grain, as thinly as you require.

Pork products

Using a meat slicer (see page 21) is the only way of producing wonderfully thin and consistently sized melt-in-the-mouth slices of cured pork products such as the many different varieties of prosciutto. However, if you are using a knife, use a long carving knife and simply try to slice as thinly as possible.

Sausage knives are commercially available to cut sausages. Similar to fruit knives, they are basically small serrated knives.

Step 3
Turn the bacon strips 90° and slice each piece into 1 cm (⅖ in.) lardon chunks. You may want to change to a cook's knife for this part, or continue to use the carving knife, as used here.

Step 4
If the meat is particularly fatty, simply trim the fat off. Place the pork fat-side down on a chopping board, insert the knife just above where the fat is located and slice along to remove.

SLICING CHORIZO

Hold a serrated knife in one hand and, with the other hand holding the chorizo steady, carve backwards and forwards to slice the chorizo. This method can be used to cut any thin salami-type meat.

fish and
seafood

There are a number of knife skills necessary for the preparation of this food group. Be it preparing live langoustines, filleting a fish, shucking oysters or shelling scallops, there is a knife to suit and a correct procedure to follow. Whatever you are preparing, a sharp knife is absolutely essential for these techniques.

Fish and seafood knives

A diverse selection of knives can be used in the preparation of fish, some of which are extremely specialized. Perhaps the most important is the fish-filleting knife. It is one of the few knives that has a flexible blade that is a perfect aid to the fish filleting process, and will become more familiar with practice.

Fish-filleting knife (flexible knife)

The perfect knife for getting around tricky fish fillets, this tool is extremely effective at releasing them from the bones. The filleting blade is almost always thin and long but the shape tends to differ according to the manufacturer. Choose that which you find most comfortable to use.

Oyster knife

A strong hand is needed to wedge into the tightly shut shell of an oyster, but the task is made easier with this chamfered specialist blade. There are now tools on the market for gripping the oyster, but a thick tea towel will work fine.

Salmon slicer

The hollowed-edge blade responds beautifully to oily fish, ensuring the flesh doesn't stick to the blade. The length of the blade enables long sweeping cuts and is ideal for slicing smoked salmon or cutting salmon into escalopes.

Small Santoku knife (with hollowed edge)

Great for cutting fine tartare of tuna or intricate fish work, the ergonomics of this short blade enable good control and precise cuts. Some knives may come with a descaler on one side.

Poultry shears

Useful when cutting through tough bones of large fish and trimming fins, gills and tails, it's handy to keep a pair nearby.

Sashimi knife

Usually bevelled on one side only, the blade of this knife is super sharp and is designed to cut through the flesh of fresh fish as efficiently as possible. A must-have for sushi chefs. This knife requires a very different cutting movement compared to other Western knives, and because it is bevelled on one side only, it may at first feel alien to hold. As with all knives, handle with care.

Other useful tools

FISH TWEEZERS

These are useful for removing fine bones from very bony fish such as red mullet (see page 21). If the bones are difficult to pull out it is a good sign as it means the fish is very fresh. If you don't have a pair of fish tweezers, a clean pair of cosmetic tweezers can be used instead.

FISH DESCALER

Not the most important tool (a fishmonger would normally do this for you) but a very efficient one nonetheless (see page 21). If a fish needs to be scaled and you don't have a descaler, try using a cook's knife at an angle, working backwards along the fish's skin (see page 142).

Flatfish

Flatfish such as halibut, sole or plaice have four fillets; two on the top and two on the bottom. One of the fillets on each side is naturally longer than the other – before you begin filleting, use your fingers to feel where each fillet ends.

Learning to skin and fillet a fish are useful knife skills. Skinning decreases cooking time and often produces very tender fillets, however, the fillets will also fall apart more easily and be slightly less flavoursome. Always make sure your surface is clean and dry before you begin skinning. For best results, use a pinch of salt at the bottom of the fish where you start to fillet – this makes it easier to grip.

WHICH KNIFE?

Filleting

The flexible fish-filleting knife is ideal for filleting, as it allows freedom of movement around the fillets. Filleting is also possible with a regular cook's knife, but perhaps not as easy.

Removing fins

Poultry shears are useful for snipping away any fins.

Pre-preparation tip

Before beginning to prepare, wipe the fish on both sides with a kitchen towel. This will prevent the fish from sliding around the board.

FILLETING A FLATFISH

Step 1
Place the fish white-side down on the chopping board. Using the belly of a knife, score a straight line down the line that lies in the middle of the fish (i.e. down the backbone). Cut all the way along, through the skin of the fish.

Step 5
As you remove more of the fillet, gradually move the knife to the side to reveal more bones. As you reach the edge of the fish the fillet will end and the 'frills' – the connecting tissue that attach the fillet to the edge of the fish – begin. Release the entire fillet from the fish, but leave it attached to the side of the fish by the gills.

Step 2

The knife will fall to one side of the backbone, hitting the bones as you proceed – making a ticking sound. Ensure your knife follows the flesh of the fillet all the way to the top of the fish. Use your fingers to feel where the fillet ends if you are not sure. Remove as much flesh as possible.

Step 3

Make a horizontal incision just above the tail, where the soft flesh ends. This will make removing the fillet easier. Simply cut through the skin, down to the bone.

Step 4

Return to the main fillet now. Using the tip of your knife, take the longer fillet away from the main fish body, working from the top of the fillet downward – you should create a ticking sound as your knife hits the bones. Use the tips of your fingers to pull back the fillet while you cut.

Step 6

Once the entire fillet has been detached from the bones (but is still attached by the gills), pull the fillet away from the fish. Steady the fish by placing the heel of your hand (use a kitchen towel if you like) on the top of the fish, preferably on the head. With the other hand, grasp the top of the fillet and pull downwards and to the side. Pull hard.

Step 7

Repeat steps 1–6 to remove the remaining three fillets.

Filleting tip

As with butchering meat, keep your knife close to the bones when filleting a fish. This ensures that you will cut as much flesh as possible away from the bones, leaving a cleaner finish. If the knife cuts into something soft or squishy, this is probably the prized flesh you are cutting into; stop immediately and retrace your steps.

SKINNING A FLATFISH FILLET

Step 1
First clear your chopping board as much as possible, wiping it free of any moisture. Place one fillet on the board, skin-side down, and sprinkle a pinch of salt on top of the fish.

Step 2
Squeeze your knife into the tiny gap between the skin and the start of the fillet. Hold the knife at an angle of approximately 25°. As soon as the skin becomes loose enough, grip it with the fingers of one hand. Using a sawing motion, cut the fillet away.

Step 3
To make the skinning process cleaner and easier, fold the fillet over whenever possible so it remains out of the way. The further up the fillet you cut, the tighter you will be able to pull the skin. Keep cutting until the fillet is free.

WHICH KNIFE?

Skinning

When skinning fish, a flexible knife really comes into its own. The main part of the blade bends onto the board, allowing the hand to be slightly above the fish, so knuckles are out of the way of the board.

Scaling

If you aren't using a special fish descaler, use a cook's knife for this task.

Skinning a whole flatfish

It may be preferable to leave the fish on the bone but remove the skin. To do this, work from the tail end of the fish. Cut across, just above the tail (as in step 3, page 141). Lift a little of the skin and pull briskly downwards, taking all the skin away.

SCALING FISH

A fishmonger would normally scale a fish for you – scaling a fish at home is a messy job – but you can buy a special fish descaler. Some modern knives have a descaler on one side, otherwise simply use a cook's knife, as shown above. Scrape along the fish at an angle. Always scrape backwards against the fish – in the opposite direction to the way the fish swims.

Roundfish

Roundfish such as salmon, mackerel and cod are identifiable by their round body and eyes located on either side of their head. They consist of two large fillets that are separated by the spinal bone.

A sharp knife is essential for preparation here, although flesh can be very soft and a sharp knife can slip easily, so beware. Always aim to leave little flesh on the bone and release the biggest, smoothest fillets.

Why do we gut fish?

Gutting can make the fish taste better and it ensures an overall cleaner end result. Fish usually arrive gutted unless you ask for them otherwise. The guts of a flatfish are never as prominent as a roundfish, and can usually be removed by hand quite easily during filleting.

GUTTING A ROUNDFISH

Step 1
Wipe the fish so it is slime free. Take the heel of one hand and place it on the fish, away from the path of the knife. Hold a fish-filleting knife with the other hand. Using only the tip of the knife (this is important to avoid disrupting the guts too much), score through the skin from the anal vent to the fins just below the gills, cutting away from you.

Step 2
With the belly open, turn the fish spine-side down. Delve into the guts and carefully pull everything out. Disposable gloves may be a welcome option at this stage.

Step 3
Most of the guts are connected to the fish at the head end. Using kitchen scissors, snip all of the contents of the guts from near the head and then discard. Wash the fish out and pat dry with a kitchen towel.

FILLETING A ROUNDFISH

Step 1
Begin filleting a roundfish by trimming the pectoral fins (as pictured here), using poultry shears or kitchen scissors. Next, remove the dorsal fin from the spine of the fish.

Step 2
Cut through the skin on a diagonal, just behind where the trimmed dorsal fin is located. Try to maximize the amount of flesh you cut away.

Step 3
Cut through the flesh until you reach the bone. Stop at this point.

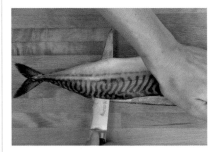

WHICH KNIFE?
Filleting and pin boning

As with filleting a flatfish, the flexible blade of a fish-filleting knife is perfect. Substitute a regular cook's knife if a bendy blade isn't available. For pin boning, fish tweezers are ideal – it is difficult to do this with anything else.

Skinning a roundfish

Skinning a roundfish is exactly the same as skinning a flatfish – remember to keep the knife at a sufficient angle in order to remove as much flesh as possible, yet not harsh enough to pierce the skin.

Step 7
Place the heel of your hand on the fatter part of the released fillet and press down to keep the fillet in place. Keeping the knife at a 30° angle, slice through the fillet until you reach the end of the tail. Remember to keep your knife moving along the spine at all times – the aim is to leave no flesh on the bones at all.

Step 8
The fillet will now be fully released. To trim the fillet, take the knife and remove the thin part of the belly flap. Using your knife at an angle, cut away from you, slicing away the unwanted flesh. Fillet the other side by simply turning the fish over and repeating steps 1–8.

Step 4
Turn the fish around and score along one side of the backbone, along the fish. Cut just through the skin, keeping your knife as close as you can to the backbone. Start by using the tip of the knife, and proceed using the belly of the knife – this will give more control.

Step 5
With the knife leaning next to one side of the spine, use long sweeping motions to remove more of the fillet. As soon as you can, pull the fillet back with your fingers – this will make it easier to see what you are doing. Your knife should scrape along the central backbone at all times.

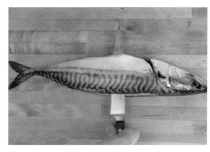

Step 6
Once you have cut the fish all the way to the tail and started to remove the fillet, insert the knife into the centre of the fish.

PIN BONING (DEBONING)

Step 1
To locate the bones, run your fingers backwards down the centre of the fish, to make the bones stand up, and so easier to pull out. Depending on the type of fish, there may be bones located at the bottom of the thicker part of the fish (such as the belly flap). To remove these, simply hook your knife underneath and pull them away.

Step 2
Use the fish tweezers to pull the bones out. Hold the fillet down in one hand and pull the bones out with the other. If the bones are stubborn in coming out – a good sign – it means the fish is very fresh. If you don't have any fish tweezers (also known as a pin boner), a pair of regular clean tweezers will work fine.

Scoring the skin
For faster cooking, or if you are cooking a fish whole, score the skin and flesh of the fish with a knife. This will also ensure even cooking, particularly in the thicker parts of the fish. This method is particularly effective when marinating fish because it allows the marinade to penetrate further into the flesh.

BUTTERFLYING SARDINES AND ANCHOVIES

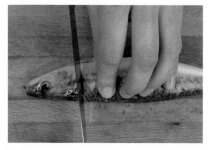

Step 1

Place a scaled and gutted sardine (or similar) on the chopping board. Grip the fish in one hand and in the other hand, take a cook's knife or Santoku knife. Remove the head by cutting directly through the fish, and discard.

Step 2

Slice through the belly by scoring all the way from the top to the bottom. Don't cut all the way through the fish, just through to the bones.

Step 3

Turn the fish cut-side down, with the cut edges of the fish turned outwards. Hold the tail and press along the backbone, running your hands up and down the fish to open it out.

Step 4

Continue pressing until the fish is flat and has been completely opened up.

Step 5

Turn the fish over. Grip the top of the fish with one hand and pull the bones away with the other hand. They should come away quite easily (particularly with sardines as the flesh is soft).

Step 6

When the bones are only attached at the bottom of the fish (by the tail), snip them at their base with a pair of scissors. Leave the tail on for presentation purposes, if you wish.

Salmon

Salmon, a roundfish with flesh orange to red in colour, is extremely versatile. It can be cut into paves (slabs), smoked, and enjoyed cooked or raw (in sashimi or nigiri).

Whatever your preference for eating salmon, there are a number of key knife skills fundamental to its preparation. In most cases, clean, precise cuts are required, so always ensure you are using the appropriate knife for the task in hand, and make sure your knife is sharp.

WHICH KNIFE?
General preparation

As salmon can be large, a large cook's knife is ideal. To portion a salmon into steaks (which involves cutting through the bone), use your most robust cook's knife or a meat cleaver or bone splitter, if available.

Portioning

To portion a fish equally, first decide on how many portions you need and then score the fish lightly into the chosen portion sizes using your cook's knife. This way you can ensure that all the pieces are as even as possible before you cut them fully. If you find you haven't shared the pieces equally, simply adjust the sizes before cutting.

CUTTING SALMON INTO STEAKS

Step 1
Place your whole scaled and gutted salmon (see pages 142–143) on the chopping board. Remove the head using a large cook's knife. This can be done in one forward motion, cutting right through the spinal bone. Discard.

Step 2
Hold the fish with one hand, with your fingers tucked out of the way. Measure the portion size you require for the steaks, and make a clean slice through the fish. Try to make this cut as even as possible.

CUTTING SALMON INTO PAVES

A pave is a French term for cutting salmon into slabs. First, fillet a salmon (see pages 144–145). Decide how many portions you would like and simply cut through the skin, flesh side first. You can also try cutting the salmon portions on a diagonal – this offers a more attractive, professional finish.

CUTTING SASHIMI

Step 1
Sashimi essentially means raw. For more details, see panel on page 149.
Take a piece of skinned salmon fillet and trim into equal, neat blocks.

Step 2
Slice into lovely sashimi-style pieces, pulling the slice towards you as you cut.

WHICH KNIFE?

Paves

A large cook's knife is perfect for cutting salmon into paves.

Sashimi and nigiri

A razor-sharp sashimi knife is designed specifically for this purpose. Alternatively, a razor-sharp Santoku knife or large cook's knife with a hollowed edge will work fine.
Note: the natural motion of the sashimi knife (because of the single bevel) is very different from Western knives – using requires practice.

CUTTING NIGIRI

Step 1
Nigiri is sushi laid over bricks of rice. Preparing nigiri requires that the fish be cut on an angle, for a better texture. To cut salmon into nigiri, begin by taking a block of salmon, and cutting off a triangular piece.

Step 2
Then, slice a piece on the horizontal – approximately 5 cm (2 in.) in length. Continue along, cutting perfect slices of salmon for nigiri. Aim for the slices to be slightly larger than the blocks of rice.

CUTTING SALMON INTO ESCALOPES

Step 1

Position the salmon so that you are cutting away from yourself. Start at the tail end of the salmon. Using the fish-filleting knife, slice at an angle of approximately 30°. Using one slice, attempt to cut an even portion. The knife should just skim the end of the fish. If a fish-filleting knife isn't available, use a cook's knife as above.

Step 2

Slice away from you, using long clean slices, continuing to slice all the way up the salmon.

TAKING SALMON OFF THE BONE

Step 1

To take poached salmon off the bone, remove the skin with the back of a knife as delicately as possible (the flesh will blemish very easily).

Step 2

Use either two spoons, a palette knife, a large spatchelor or the back of a large knife to then remove the salmon flesh in large pieces, working along its natural lines.

Salmon sashimi

Sashimi means 'raw'. The most important point to make about cutting any fish into sashimi is to ensure that the fish is sushi grade (i.e. as fresh as possible). Given the years of training that sushi chefs have to undertake in order to work with fish, the instruction on this page should be considered as guidance for how to prepare premium fresh raw fish. Fish for sushi and sashimi may be cut straight down, angled, flat or wafer-thin – there are various benefits and downfalls to each of these. Illustrated here is the straight down cutting method. But before you begin, make sure your knife is razor-sharp – as with any knife, remember to handle with care.

Smoked salmon

Smoked salmon is either sliced into escalopes (thin, boneless, round-shaped slices) or into very long, elongated pieces. A very long, thin knife is ideal because it is long enough to make one clean slice in one direction – if you don't have access to a long, thin knife, use a cook's knife.

Monkfish

Monkfish has a unique shape: a rather oversized monstrous head that overshadows the meaty tail. The only large bone is located where the two fillets lie. Once the sinewy skin is removed, monkfish is then remarkably easy to prepare.

Removing this skin can be tricky, however. Try to take the minimum amount of flesh away. Monkfish don't have scales so this is one less thing to worry about. Deliciously dense in consistency, the monkfish marries well to most flavours. If possible, use a flexible fish knife to prepare this fish.

WHICH KNIFE?
Filleting and slicing

For filleting, the flexible blade of a fish-filleting knife is ideal. A medium to large cook's knife will also work well, however, as the fillets are attached to the bone in a very straightforward way.

For carpaccio, use a sashimi, Santoku or cook's knife.

Trimming spines

To trim the dorsal spines, use poultry shears or kitchen scissors.

Carpaccio of monkfish

With fresh monkfish, why not try a carpaccio? Simply prepare into fillets and freeze for an hour. Then slice as thinly as you can with your sharpest sashimi, Santoku or cook's knife, as you would with tuna (see page 153). Serve with olive oil, lemon and salt and pepper.

FILLETING AND SKINNING

Step 1
First, trim the dorsal spines using some poultry shears or kitchen scissors. Pull the spines upwards with your fingers and snip them at the base. This will make the removal of skin easier. Take care as the spines are sharp.

Step 5
Next, start to remove the fillets. Place the fillet in front of you. With the knife positioned to one side of the central bone, cut downwards, keeping your knife as close to the bone as possible. Hold back the fillet that is being released. Remember to keep your knife close to the bone throughout.

Step 2
Remove the sinewy skin. This is totally indigestible and easy to identify because it is dark in comparison to the colour of the rest of the fish. Lift up the skin and cut it back with a cook's knife. Repeat on both sides until all the skin, from the fatter part of the fillet to the middle, is released. Try not to cut into any of the flesh.

Step 3
Use the heel of your hand to secure the top of the fish. With the other hand, briskly pull the skin backwards. It will gradually come away.

Step 4
Continue to pull the skin all the way off. It should easily slip over the tail.

SLICING MONKFISH FOR CEVICHE

Step 6
Continue cutting along the fish bone until the entire fillet has been released from the fish. Repeat to the other side so that you are left with two long fillets of monkfish. Trim the thin membrane from the fillets, retaining the natural shape of the fillet as much as possible.

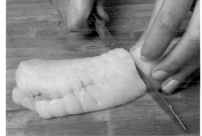

Step 7
In order to roast an entire portion of monkfish, simply slice and cut away the required size of pieces.

Ceviche, the South American method of 'cooking' fish in lemon or lime, is either prepared in dice or slices. A beautifully refreshing summer dish, lots of fish are ideal for this method of preparation. If you decide to use slices, slice the fillet thinly into equal strips as shown above. For diced ceviche, turn the cut strips 90° and cut again.

Tuna

Fantastically versatile, tuna combines well with a variety of flavours. Tinned, as tartare or pan-fried rare – tuna is delicious.

A great way to enjoy fresh tuna is to cut it into carpaccio (tuna sliced very thinly). You can also use the mincing technique using two knives in a double action, shown in the meat section on page 120. However, hand cutting small, intricate dice (tartare) is a fantastic way to show off newly acquired knife skills. Ideally, tartare should consist of perfect squares of no more than 5 mm (⅕ in.) squared. Whatever your method, always prepare the tuna straight from the fridge to keep it as firm as possible, unless your recipe requires otherwise.

TUNA TARTARE

Step 1

Rather like blocking-off a carrot (see page 47), begin by blocking-off the tuna. Try to keep the edges straight at all times as this will ensure full use of the piece of tuna, and result in uniformly sized pieces. This can be difficult as tuna has a naturally squashy consistency.

Step 5

Turn your chopping board 90° and cut the pieces further into 5 mm (⅕ in.) squared cubes.

WHICH KNIFE?
General preparation

As tuna is an oily fish, it responds very well to the hollowed-edge blade of a Santoku or cook's knife. A sashimi knife will slip through the soft flesh very easily too.

Carving carpaccio

For carving thin carpaccio, a hollowed-edge carving knife is ideal, otherwise use a very sharp cook's knife.

Tuna tips

Slicing is easier (and knife control more effective) if you freeze the tuna slightly first, until it is firm but not frozen. If you have a meat slicer (see page 21), use it on this occasion, otherwise use a long carving knife or a knife with a long blade.

Like preparing avocado, handle tuna delicately and as little as possible, otherwise it will start to break down and spoil.

Step 2
Cut the tuna into slices measuring no more than 5 mm (⅕ in.) in thickness.

Step 3
Place the slices flat on the chopping board, one piece at a time. Cut into long thin slices, no wider than 5 mm (⅕ in.).

Step 4
Continue until the whole piece is cut into thin, even strips.

TUNA CARPACCIO

Using the claw method to grip the tuna (see page 32), use a very sharp knife to slice downwards, cutting the tuna into perfect even slices. Be as accurate and consistent as possible.

SLICING TUNA INTO STEAKS

Take a tuna loin and place on the longest, flattest side so that the natural lines of the fish are running horizontally. Take a large cook's knife, and using the claw method to grip the tuna, cut into steaks of around 2.5 cm (1 in.) in width.

SLICING A COOKED STEAK

Secure the tuna steak with tongs. Take a cook's knife and slice downwards into portions measuring around 2.5 cm (1 in.) in width.

Oysters

Shucking is really the only preparation you will have to do with oysters. To shuck an oyster simply means to open it up and leave it in its shell. Oysters are difficult to open, however.

An oyster knife is designed so you can grip the knife securely and, using the pointy tip, penetrate the small gap in the oyster. Oyster knives are essential for this practice (and conveniently, they are very cheap). Once the tip of the knife is inside the oyster, at its narrowest point, it should be used in an up-and-down lever motion to unhinge the shell of the oyster.

WHICH KNIFE?
Shucking

The oyster knife is ideal for this task and comes in many shapes and sizes. For the most efficient use, look for a knife that has a very thin tip and a short, stubby blade, to squeeze into the small gap quicker.

Oyster knives are incredibly cheap to buy and essential for shucking. If you don't want to invest in an oyster knife, you can buy oysters already shucked – however the juice can spill out of these, making them dry out.

Safety tip

Opening an oyster is a tough job. A lot of pressure has to be put onto the knife as the shells are very tightly shut, so take care when shucking – the knife can very easily slip and cause injury. Gloves – either specialized chain mail or rubber/plastic – are a good precaution to protect hands. In the sequence opposite, we have used a tea towel in the process. If using a towel, make sure you use a double thickness for extra safety.

SHUCKING OYSTERS

Step 1
Wrap your hand in a tea towel – use double thickness to be extra safe. Make sure no part of the hand is exposed. Leaning on a chopping board, grip the oyster with this same hand. Hold the oyster curved-side down and look for a small gap at the pointy end of the shell.

Step 5
Start to pull the lid back from the oyster, gently releasing the gills of the oyster from the lid. Put the lid to one side.

Step 2
Wedge the tip of the oyster knife into the tiny gap in the oyster. Really dig in at this stage – this will enable the knife to move up and down and lever the oyster open. If it isn't wedged in well enough at this point, you run the risk of crumbling the oyster shell away instead of entering into the hole.

Step 3
Once the oyster starts to open, move the knife around the edge of the oyster. The main oyster sits at the bottom of the curved shell but is attached to the lid of the oyster shell by its gills. Be careful not to hastily rip open the top of the oyster, otherwise it may tear.

Step 4
Slide the knife all the way around the edge of the shell, being careful not to rip open the oyster, just aiming to open the shell.

Step 6
Use the knife again to gently loosen the oyster from the base, being careful not to pierce the oyster. This will ensure that the oyster will slip easily down the throat once it's ready to eat. Try also to retain the natural juices – be careful that they don't spill out.

Step 7
Once the oyster is released from the base, use your knife to turn the oyster over, for the best presentation.

Step 8
For tempura or other uses without the shell, simply tip the oyster and juices out of the shell and use accordingly.

Scallops

Scallops can be a challenge to shell perfectly as it is impossible to see inside the shell at first. The meat is attached to both sides of the shell, making it easy to cut into the meat or tear it while opening.

In order to open the scallop safely, partially open it and then immediately cut the meat away from the flat side of the shell before proceeding, to avoid damaging the meat. Very fresh scallops will still move (pulse) slightly while you are shelling them, so beware! A flexible fish-filleting knife (or a cook's knife) is ideal to use when preparing scallops.

WHICH KNIFE?
Shelling scallops

Ideally you should use a flexible fish knife to open a scallop shell. The flexible blade allows the required freedom of movement. The task can be completed with a regular cook's knife however, or any knife that is long and relatively thin.

Benefits of buying scallops in the shell

Learning to shell scallops is a great skill to learn, as by buying scallops whole you can be sure of their quality. Scallops sold loose have been known to be soaked in water to add extra weight. Also, as preparation is being done at home, it is cheaper to buy scallops in the shell rather than without. The shell can also be used for Coquilles St. Jacques (see Glossary, pages 172–173) or similar.

SHELLING/REMOVING THE MEAT

Step 1
Place the scallop flat-side down on the chopping board and place one hand on top of the curved shell. Carefully insert the tip of the knife into one corner of the shell and start to wiggle the knife gently to partially open the shell. Do not open the shell fully here as you risk tearing the meat.

Step 5
Carefully remove the entire contents of the scallop with a spoon. The meat will be slightly attached to the curved side of the shell, although not as much as the flat side. Gently tuck the spoon behind the meat and pull the meat away – everything else (gills, etc.) will follow. Leave no remains in the shell. Set the shell aside.

Step 2

Move the point of the knife just around the edge of the curved shell. This will start to open up the shell. The knife will be in contact with the gills – this is fine.

Step 3

As soon as the shell starts to open, take the knife and, keeping it close to the flat side of the shell, make a 90° turn. Then, using your knife at a slight angle (30°), scrape the shell as you go, cutting and releasing the scallop from the flat side of the shell.

Step 4

The shells should pull apart easily now. All the shell's contents will be contained in the curved side of the shell, leaving little or no remains on the flat side.

Step 6

Remove everything surrounding the scallop meat (the gills, the black stomach sack and the coral if there is any). Begin by placing your thumb between the black stomach sack and the meat. Be sure to protect the scallop meat during this process as it can tear easily.

Step 7

Pull away the membrane that surrounds the scallop meat with your fingers.

Step 8

The meat should now be free of all membranes. Wipe the scallop clean with a kitchen towel. The gills can be used for stock. Wash the shells for presentation, if desired.

Prawns

An incredibly popular shellfish due to their lovely flavour, prawns are very well suited to freezing, which makes them easy to keep. They are relatively easy to prepare – most of the preparation lies in the peeling.

A more intricate knife skill to master is the technique of butterflying. Butterflying a prawn involves opening it up in two equal matching parts – creating a shape similar to that of a butterfly. Butterflying prawns, like butterflying lamb, ensures quicker cooking and a different presentation. Marinades are also more thoroughly distributed when the butterflying technique is used as a wider surface area is created.

Step 1
Remove the prawn head first. Hold the prawn by the thickest part of the tail in one hand and hold the head with the other hand. Simultaneously twist both parts, and pull apart.

WHICH KNIFE?
General preparation

To remove the intestinal track, use a kitchen knife or turning knife – the short blade of these knives offer plenty of control for this fiddly task.

For butterflying, use a small cook's knife or small Santoku blade to aid precise cutting.

Fingers are best used for shelling.

Butterflying shelled prawns

Shelled prawns can also be butterflied. Simply cut a slightly shorter length into the prawn than when preparing the unshelled prawn (see step 1 of *Butterflying Prawns* on page 159).

Step 4
The meat should now be released, and completely intact.

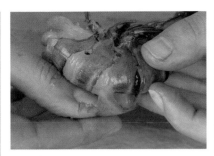

Step 2
To remove the shell, dig your finger underneath the shell, precisely where the shell ends. Grip the tip of the shell and pull, taking one segment of shell away at a time.

Step 3
Once you've reached the tail and all of the shell has been removed, pinch or squeeze the tail end, right at the end of the tail meat. This should cause the end of the shell to come away easily.

To shell or not to shell?

Cooking prawns with their shells on intensifies the flavour and colour. Shelling prawns while eating is also a sociable way of eating and means food is digested in the ideal way: slowly. Shelled prawns, on the other hand, are quicker to cook and easier and cleaner to eat. For the best of both worlds – and for more flavoUr – fry prawn heads and the shells in the frying pan with the shelled prawn, this way colour and flavour are obtained with little fuss.

Peeling prawns also means you can remove the black intestinal tract. Sometimes this isn't visible at all, in which case, leave it in.

Step 5
To finish, locate the dark intestinal tract, which is clear and sometimes disguised. Dig just underneath the tract with a turning knife or kitchen knife and begin pulling a little out. Proceed by extracting the entire tract, and discard.

BUTTERFLYING SHRIMP

Step 1
Remove the head of the unpeeled prawn then cut into the flesh of the tail. Cut through the shell, through the length of the body but leave the tail end intact. The prawn will now create a 'Y' shape.

Step 2
Turn the prawn over and with your fingers, press the main body of the prawn out, so it is fully spread but still attached by the tail. This is similar to the technique used for butterflying sardines (see page 146).

Lobster

Lobsters are best eaten after having been cooked live, because their flesh hasn't started to deteriorate. Preparing lobster is therefore not for the faint-hearted. That being said, never buy a dead lobster – like all fish, they should be as fresh as possible.

There are two ways to prepare a live lobster. Either freeze the lobster for two hours and then plunge it into boiling water – this method is ideal if you want to keep the main body of flesh in one piece – otherwise, as illustrated here, cut the lobster in half lengthways while it is still alive. This is an ideal method if you want to cook the halves quickly under a grill. Either method requires you to be precise and definite with your knife.

WHICH KNIFE?
General preparation

Use a large cook's knife or a meat cleaver or bone splitter to cut through the lobster's shell and divide the body.
   To remove the intestinal tract, use a small knife such as a turning knife or kitchen knife.

Handling a live lobster

Pick up a lobster by holding it just behind its head on the main body. When being held, lobsters will naturally arch their back and spread their claws out. Keep your hands just behind their head to avoid any contact. Kitchen tongs minimize contact when picking them up, but aren't ideal as the shells are shiny and thus difficult to grip. Lobster claws are normally bound with an elastic band so they do not harm each other (or you).

PREPARING A LIVE LOBSTER

Step 1
Place the lobster on a chopping board. Ideally it should be alive and kicking – to calm it down, rub the head backwards and forwards.

Step 5
The lobster should now be in two equal halves. The halves may still be moving slightly – this is normal.

Step 2
Take a large cook's knife and with the tip, insert the knife into the main body of the lobster.

Step 3
Move the tip of the knife quickly to the bottom of the lobster, then in a lever-like action, cut downwards 90° to the board, to cut the whole head in half, so your knife is resting on the board.

Step 4
Turn the chopping board or the lobster around 180°. Insert the tip of the knife through the shell until you reach the chopping board then, keeping the tip of the knife in one place, bring the knife down in one precise lever action to cut through the entire body.

Step 6
Take one of the halves of lobster and remove the stomach sack from the tail section using a spoon. Do the same with the other half.

Step 7
Using the point of a small knife (such as a turning or kitchen knife), remove the intestinal tract from both shells.

Step 8
To crack open the claws, use a meat tenderizer or a rolling pin covered in cling film, or the flat side of the heaviest knife you have. With a few assertive motions, crack the claws so that two or three large cracks appear. This ensures that the claw meat cooks at the same rate as the main body of the meat. It is now ready to cook.

Squid

Squid can grow up to 13 metres (43 feet) in length, although the squid used in the demonstration here is a lot smaller. Squid can be reasonably quick to prepare, especially if the squid ink sack isn't pierced.

Squid is most commonly cut into pieces, or rings – known as calamari. Scoring squid – cutting it into flat pieces and then scoring the inner side – produces a curled-up roll of squid, which looks great and tastes fabulous stir-fried with chilli salt. Scoring squid is a great way to gain good knife skills as it requires precision and consistency.

PREPARING THE SQUID

Step 1
Pull the main body out of the head. Grip the head in one hand and the tentacles in the other hand and gently pull apart. The intestines should come out at this stage too. The ink sack, identifiable by its small, pearly white pouch, is located near the intestines.

SCORING SQUID

Step 1
Cut the tube of squid along its natural seam. Open out the squid.

WHICH KNIFE?
General preparation

For general squid preparation, a regular cook's knife works well. A flexible blade isn't really needed but it can be used.

For scoring squid, the shorter fatter blade of a Santoku knife offers perhaps more precision than a cook's knife, but again a medium cook's knife is fine.

Squid rings

To cut a squid into rings, take a cleaned main body of squid and, using a medium cook's knife, simply slice into rings of desired thickness.

Step 2
Slice the tentacles just below the head, in front of the eyes. Discard the remaining head.

Step 3
Pull out the plastic quills from each side of the squid, along the seam. These will come away easily.

Step 4
Pull away the two fins from either side of the body pouch. The maroon and clear skin should come away from the body at this stage. Do eat the delicious fins – their slightly firmer texture is a result of having been used more than the squid's main body.

Step 2
Using your knife at an angle, scrape away the jelly-like substance from the inside of the squid.

Step 3
Cut the squid in half again lengthways (cut smaller if required).

Step 4
Take a piece of squid and score the flesh in diamond patterns, but be careful not to cut right through the squid.

bread,

and cake

pastries

Whether making croutons or cutting beautifully prepared pastries or sticky cakes, these foods need to be handled with care, and so using the appropriate knife and knife skills is crucial. For foods in this section, a wooden bread board provides an ideal working surface.

Bread, pastry and cake knives

This is where the serrated blade really comes into play: it is perfect for cutting through tough crusts and delicate interiors. The scalloped edge (which mirrors the indentations of the serrated blade) should be used for cutting through delicate cooked pastries.

Patisserie knife
(Also known as a confectioner's knife or pastry knife)
Very long and sturdy, these knives are extremely useful in pastry preparation. The scalloped/wavy edge ensures delicate yet precise cutting through numerous pastry layers.

Serrated bread knife
Ideal for carving bread of all kinds, whether slicing or making croutons, this is the only knife that tackles the crustiest of crusts and softest of centres effectively.

Cook's knife
The versatile cook's knife has a role to play with all food groups.
Heat it up and it will cut through the stickiest of cakes easily
(see page 170).

Small serrated knife
A great knife for making small croutons or carving bread when a large
serrated bread knife is too big and greater control is required.

Palette knife
When lifting delicate pastries, or slices of sticky cake, the bendy
(but blunt) blade of this knife is great as its flexibility enables it to
manoeuvre around difficult situations.

Bread

The main knife skill required for preparing bread is that of carving or slicing, although there are useful knife skills to be learnt for making croutons and Melba toast.

Croutons can be made from almost any type of bread – whether in a loaf, sliced, with or without a crust or from a baguette. Melba toast is best made from pre-sliced bread, and requires precision and a suitable knife. Always ensure you use a serrated bread knife for both of these tasks. The indentations of this knife are specially designed to tackle tough crusts without damaging the bread's more delicate interior.

CROUTONS WITH SLICED BREAD

Step 1
Before you begin, decide how thick you want your particular croutons to be. Cut off the crusty edge with your serrated knife.

MELBA TOAST

Step 1
A great accompaniment to pâté, Melba toast is easily made with sliced bread. Begin by cutting the crusts off the slices of bread as shown.

Step 2
Toast the crustless bread and then halve the thickness of the bread by cutting in between the two toasted faces. This is easily done, as the crusty exterior of the bread deters the knife from coming through. Cut all the way across.

Step 2
Cut the bread into even fingers, keeping the thickness the same as the height.

Step 3
Take two fingers of bread at a time and turn them 90°. Start to cut them into cubes, matching the thickness to the height and width, for perfect crouton cubes.

Step 3
Slice the bread diagonally into triangles.

Step 4
Now toast these slices on the un-toasted side. Your Melba toast is now ready.

Croutons made with baguette

The same crouton-making method used with sliced bread can be applied to baguettes. Provided you use a small baguette, these croutons are a perfect bite size.

Take a baguette and lay on a large bread board or chopping board. Carve the loaf as thinly as you require (1 cm [⅖ in.] or less). A sharp serrated knife is crucial, otherwise the loaf will squash and the shape will be distorted.

For more elegant-looking croutons, cut on the diagonal.

Pastries and cake

Pastries (both raw and cooked) and cakes can pose their own cutting difficulties, but there are a few knife techniques to facilitate the tasks.

It is essential to use a sharp knife when cutting raw pastry – if the knife drags, the fine pastry layers will stick together and not rise properly. The knife must also be floured to ease cutting. For best results, cut the pastry straight from the fridge.

With cooked pastries, use a sharp knife and remember to be confident and assertive with your cutting technique. To cut a perfect slice of sticky cake, heat your knife in boiling water prior to use.

CUTTING VERY STICKY CAKES

Step 1
Submerge the knife in very hot water for a good five minutes (simply boil the kettle a few times and pour the water into a jug). It is essential to get the blade very hot. Remove the knife from the water and wipe the blade dry.

CUTTING RAW LAYERED PASTRIES

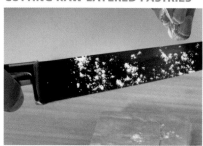

Step 1
First, flour your chosen sharp knife well.

Step 2

Place the knife in the centre of the cake. Support the tip of the knife with one hand and press down firmly. The knife should melt the cake slightly and glide right through it.

Step 3

Once the knife reaches the bottom of the cake, remove it. At this point wipe the knife clean and reheat in hot water, before making your next cut.

Step 4

Continue cutting until the whole cake is cut.

Step 2

Using a definite downward cut, slice all the way through the raw pastry. You may find it easier to use one hand to balance the tip of the knife, with the other hand on the handle as usual. Cut straight downwards – do not drag the knife.

CUTTING COOKED PASTRIES

If the pastry is small enough, use the entire knife and slice downwards. Simply place one hand on the handle and the other on the tip of the knife and cut straight down.

CUTTING A SPONGE CAKE IN HALF

This technique is usually used to cut a cake horizontally in half, when you want to apply cream or jam to the centre. Place one hand flat on top of the cake – this helps to keep everything balanced. With the other hand holding the knife, position the knife beside your hand in the middle of the cake. Cut through slowly and carefully to the other side, trying to keep the knife straight.

Glossary

Acidulated water
Water with added lemon juice or vinegar, used to prevent the oxidisation/discolouration of cut ingredients.

Allyl sulphide
The gaseous substance released when members of the onion family are cut/chopped.

Belly of the knife
See *Cutting edge*.

Bevelled
The inclination that one line or surface makes with another when not at right angles. In Japan, knives are bevelled on one side only.

Blade
The main body of the knife (see page 10). This can be flexible or rigid.

Blocking-off
The process of removing the curved edges from an ingredient so that the remaining shape is regular (usually an oblong), from which uniform shapes can then be cut.

Bolster
The thick piece of metal between the handle and the base that provides weight and balance.

Brunoise
The method of cutting ingredients into small dice. The term can also be applied to a group of diced vegetables.

Butterflying
The process of removing the bone or shell from an ingredient and opening it out equally like a 'butterfly'. The process is often applied to a leg of lamb or prawn.

Carpaccio
An Italian term originally for thinly sliced, raw beef, although the term is now loosely applied to describe any thinly sliced raw ingredient.

Chinned
Describes a piece of meat on which the bones that connect the ribs to the backbone, or the backbone itself (chine bone) is cut.

Concasse
From the French 'concasser', to crush or grind, a cooking term meaning to roughly chop an ingredient, usually a vegetable. The term is most often applied to a tomato when it has been peeled, deseeded and chopped into neat squares.

Crown of chicken
The breasts of chicken left on the bone, with the remaining carcass cut away.

Cutting edge
The most important part of the knife, the cutting edge is just that (see page 10). This is the part that is sharpened. Also called *Belly of the knife*.

Degorging
A term used to describe the process of sprinkling a vegetable with salt to draw out excess moisture.

Dorsal fin
The fin in the centre of a roundfish, located on the backbone, behind the head.

English trimmed rack
A rack of lamb of which some parts of the bones are trimmed, the chine bone is removed, and a little fat is trimmed off.

Escalope
A thin slice of meat or fish, that can be grilled or fried.

False fillet
A small fillet connected to the main fillet and almost identical to it. False fillets are present on most meats including chicken, duck, lamb and beef.

Filleting
The process of removing the fillets and bones from a piece of fish or meat.

Flatfish
Fish which dwell on the ocean bed, with eyes located on the top of the head.

French trimmed rack
A rack of lamb of which most of the flesh and fat and all of the bones have been trimmed. As with the English trimmed rack, the chine bone is also removed.

Frills
The tissue that connects the fillet to the outside of the fish, usually present on some flatfish and shellfish. It is edible, sometimes used for stock, but usually discarded.

Grain
The arrangement or direction of meat fibres.

Guards of honour
Two French trimmed racks of lamb, arranged so that the bones cross over each other, like guards' swords.

Heel
Refers to the base of the blade (see page 11).

Julienne
A French term, to cut into long, thin strips.

Lardons
Small pieces of bacon, about 1 cm (⅖ in.) in width.

Marbling
The pattern of fat running through the flesh of meat, usually present in beef – the white of the fat combined with the deep red of the flesh, looks like a marble effect.

Medallions
Neat slices of meat (usually beef) around 1 cm (⅖ in.) in thickness, and generally cut from the fillet.

Noisette
French for hazelnut, the term is also used to describe a tender, round cut of meat, usually lamb.

Oblique
Neither perpendicular nor parallel to a given line or surface, the term is usually used to refer to the type of cut made to an ingredient. Ingredient pieces cut on the oblique, have no right angle.

Oxidise
To subject to or combine with oxygen. In the context of food, the term is usually applied to fruit and vegetables, that, after prolonged exposure to air, lose freshness and often darken in colour.

The oysters
Two pockets of delicious dark chicken meat that are attached to the chicken carcass and located at the top of the chicken thighs.

Paddywhack
The creamy coloured, inedible neck tendon present in sheep and cattle.

Paring
To remove by cutting; to cut off the outer coating.

Pave
French for slab or block, the term is often applied to fish (a pave of salmon) or sometimes beef.

Pectoral fins
The fins on a roundfish located just behind the gills on both sides.

Petit brunoise
As 'brunoise', but tiny.

Pin boning
The method of removing small bones from a fish, usually using fish tweezers.

Pith
The bitter white membrane that separates the skin and flesh of a citrus fruit.

Pope's nose
This describes the fleshy part of the tail on a chicken. Also called *parson's nose.*

Rivets
Only present on some knives, the rivets secure the blade to the handle (see page 11).

Roundfish
Fish that are 'round' in form, with a central backbone and eyes and gills on either side of their bodies.

Sashimi
A Japanese dish consisting of very thin, bite-sized slices of fresh, raw fish.

Scoring
The process of making a shallow incision. Scoring the skin on a duck or pork belly aids the rendering of the fat beneath the skin's surface.

Shucking
Shucking an oyster simply means to open it.

Sinew
The indigestible, usually silvery strips of membrane found on meat, and removed during trimming.

Tang
The upper end of the blade that fits into the handle (see page 11).

Tartare
Usually a preparation of finely chopped or ground raw meat or fish which is highly seasoned.

Tempura
Traditionally, tempura describes a Japanese dish of prawns or vegetables, deep-fried in a light batter.

Tenderloin
A term for the most tender part of most meat, also called fillet.

Tunnel boning
The process of removing a bone, without opening up the meat.

Turning vegetables
A basic culinary technique of cutting vegetables into barrel shapes.

Index

Credits

Author acknowledgments

I would like to dedicate this book to the man with the best knife skills, my dad Stephen Morris Lumb (22.10.40–07.06.92), an amazing butcher and father.

Warmest thanks go to Tony and Richard at Joseph Joseph for the chopping boards; Jayesh Patel and Miranda Johns at The Japanese Knife Company for sharing their knowledge; Paul Shelley and Marc Kinsey at Haus Marketing and Distribution for the informative knife chats; and all at Divertimenti.

Thanks also to Brian Randall and Matthew Smith at Randalls Butchers, The Ginger Pig butchers, the Michanicou brothers and Alan Solis and Frederick Lindfors at The Fish Shop, Kensington Place, for sourcing the best ingredients and Ramiro Fernandes de Margalhes and Carlos Goncalves at The Fish Shop for their exquisite knife skills.

Special thanks to my friends and family for their support and encouragement, especially Lady Caroline Wellesley, family Bullock-Webster and Iain Abrahams.

Lastly, thanks to Andy and Andy at Andrew Atkinson Photography Ltd, Frances Mcloughlin and Laura Palmer for their assistance, and all at Quarto Publishing, including Kate Kirby and Emma Clayton, and in particular, Emma Poulter, who held my hand throughout the entire overwhelming process.

Picture credits

Quarto would like to thank the following for supplying the photographs reproduced in this book:

- **AlaCook** (www.alacook.co.uk): 18b, 20m, 21b
- **Aleksei Potov/Shutterstock**: 8br
- **Carlos Caetano/Shutterstock**: 8bl
- **Cookware.com**: 18t, 18r, 19bl, 19br, 19tr, 21t, 21m (fish tweezers)

All other images are the copyright of Quarto Publishing plc. While every effort has been made to credit the contributors, Quarto would like to apologise should there have been any omissions or errors, and would be pleased to make the appropriate correction for future editions.